FLASHBACKS NO. 18

The Flashback series is sponsored by the
European Ethnological Research Centre,
c/o the National Museums of Scotland,
Chambers Street, Edinburgh EH1 1JF.

General Editor: Alexander Fenton

Other titles in the Flashback series include:

SALT HERRING
ON SATURDAY

THE FISHERTOWN
OF NAIRN LAST CENTURY

Margaret M. Bochel

TUCKWELL PRESS
in association with

The European Ethnological Research Centre

The National Museums of Scotland

For Johnny and in memory
of Alex

First published in Great Britain in 2004 by
Tuckwell Press
The Mill House
Phantassie
East Linton
East Lothian, Scotland

ISBN 1 86232 246 5

British Library Cataloguing-in-Publication Data
A catalogue record of this book is available
on request from the British Library

Typeset by Hewer Text Ltd, Edinburgh
Printed and bound by Bell & Bain Ltd, Glasgow

CONTENTS

ILLUSTRATIONS

illustrations

I

INTRODUCTION

Nairn is a coastal town at the mouth of the river of that name, hence its old name of Invernairn. It lies on the south side of the Moray Firth, some nine miles or so west of Forres. The derivation of 'Nairn' is generally accepted to be from the Gaelic 'Uisge-Nearn', meaning river of alders, and its antiquity is evidenced by written references going back to the ninth century. The good-quality rich loam soil to the south of the town coupled with ready access to the sea resulted in a duality in its economic history: farming and fishing. Its population comprised Saxon-speaking fishermen and Gaelic-speaking farmers. Nairn was the justification of the apocryphal story (repeated in the *Ordnance Gazetteer, Scotland* and elsewhere) of James IV's retort when teased by the English about the insignificance of Scottish burghs relative to the English. He had a town in Scotland 'sae lang that the inhabitants of the one end did not understand the language spoken at the other'.

Nairn was viewed traditionally as a border area between Highlands and Lowlands. For example, around 170 years after James IV's proud boast to his new subjects, Dr Johnson recorded, 'At Nairn we may fix the verge of the Highlands; for here I first saw peat fires, and first heard the Erse language'.

The attractiveness of the climate and seaside location

I

earned Nairn the nickname of 'The Brighton of the North'. Perhaps this title was pushing a point, but nevertheless it demonstrated the attraction of the town in the nineteenth century to London physicians as a suitable place for patients to recover in a dry and bracing but moderate climate.

2

ABOUT NAIRN
THE COMMON GOOD FUND

Nairn always had a problem keeping its Tolbooth decent. Old minutes show that councils had a continual struggle finding the money to keep it secure and usable.[1] As well as being a jail, it was also the meeting place of the Council and Burgh Court. In 1702, the Tolbooth was given a new lock and key and the next year some essential repairs with timber from Rothiemurchus. It was burned down by His Majesty's forces (George I) in 1716 and the rebuilding was not begun until 1734.

Nairn was a poor place, with a tiny Common Good Fund bringing in only about £50 a year. The Common Good, that mysterious fund, about which, I believe, very little is generally known, originally arose from a grant of lands or other privileges by the Sovereign for the common purposes of a Royal Burgh. Most of the objects for which the Common Good was once needed are now financed by the Council Tax, so Nairn, along with other Royal Burghs, is fortunate to have this extra money to disburse on behalf of its community. There is still a steady income from the rents of Common Good land and properties, and the trustees, members of the

1 Nairn Town Council Minutes, eighteenth to twentieth centuries.

Council, are the only people with power to decide on the best use of the Fund.

This was the Council's only source of revenue for administering the affairs of the Burgh, of which the Court and the jail were the most important. When the Council failed, as it often did, to pay its taxes to the Government, troops were quartered on the inhabitants. Because of its extreme poverty during the eighteenth century, Nairn suffered this fate for years on end – 36 years altogether. The meagre resources of the townsfolk were totally exhausted during the years 1715-1751, because, in addition to quartering for tax deficiencies, the two Jacobite uprisings in 1715 and 1745 brought many more soldiers into the town.

In August 1746, in the aftermath of the Battle of Culloden, four companies of the Royal Rifles were lodged in Nairn for winter quarters. In addition to compulsory billeting, demands were made for 'Coal, candle and blanket' for the guards and for the sick in hospital. Eventually everything possible had been squeezed out of the townspeople and the Council for help with bedclothes. The Tolbooth had by this time been partly rebuilt but the steeple was not completed until 1758 and the bills for that still not paid in 1779.

On 3 July 1751 the sad state of Nairn was recorded in a memorial to the Convention of Royal Burghs:

> Several of the Treasurers of this burgh have broke and gone off with the Town's revenue, small as it is, without accounting, for several years. Because of the smallness of the Common Good the burgh is not able to send a commissioner to the Convention to represent their grievances. The King's House, Public Halls and harbour are all decayed, so are a great part of the buildings within the Burgh. It is therefore humbly hoped

that the Convention will give them such assistance and relief
as their poor and indigent circumstances call for.

The Convention sent a deputation from the neighbouring
authorities of Elgin, Forres and Inverness to enquire into the
state of Nairn. They reported that 'Everything was as described in the Memorial, that all the Public Good or revenue
of the town consists of shore dues, one public mercat and a
calf-ward, which three articles have been toll annually by
public roup'. They also found the Tolbooth to be 'insufficient
and wants reparation'.

In 1816 the Provost pronounced the jail insecure as a prison
and a new building was planned. This time the County and
the Burgh were to share the cost and the Provost hoped that
'Such a form would be given to the new building as would
render it highly ornamental and creditable to the town'.
Although the Councillors, in their discussions about the
building, always referred to it as 'the new jail', it was to
contain, just as in former days, a court room and the Council
room, as well as cells. This is the building which still stands
on the High Street and houses the Town Clock.

The builder was my great-grand-uncle John Wilson, later
Provost Wilson of Nairn for thirteen years from 1839 to 1852.
He also built Wilson Street, the Shambles (slaughterhouse) at
Merryton and 27 Union Street, home of members of my
family for four generations. On September 27, 1817, the
Council decided to remove 'The present shambles' to make
way for the extension to the proposed jail and Court rooms.
Two weeks later they had accepted John Wilson's offer 'for
executing the mason work of the new shambles at thirteen
pounds sterling'. This was the building demolished in the
1970s by Frasers the Builders at their Merryton yard.

The new jail was completed in 1818 at a total cost of £1,391 15s 9½d. John Wilson's bill for mason work came to £190 7s 9d which was paid in full. But alas for poor Nairn, all its badly needed and carefully planned projects – the new Shambles, the new jail and the Harbour, all going on at about the same time – finally broke the bank and the Councillors were in despair. But money worries still dogged the Town Council. The street dung was fetching only 4 shillings per annum, and the Treasurer's balance in 1819 was only £5. 9s.1¾d. Next year the street dung just had to bring in more. It went up to £1. The Town Councillors were always complaining too that the County did not contribute enough to the jail because 'The inhabitants of the landward parts of the County notoriously use the jail more than the Burgh inhabitants'.

I wondered why there was nothing in the Minutes recording an opening ceremony for their fine new premises on the High Street. Perhaps they had not the heart to make any fuss about it with such large debts on their minds. The Minutes of 1823 show the Council to be obsessed with the need to get those debts paid off, and soon they were disposing of their assets, the Common Good Lands. They knew quite well that even with the money raised in this way they would have a diminishing return on what was left, but there seemed no other way out. The building of Nairn Harbour and the Town House made Nairn bankrupt in the early years of the nineteenth century. Lands at Whinnieknowe, The Crook and The Links were disposed of to pay the debts, but Nairn now had a fine Town House, a modern shambles and a good safe harbour built by the famous engineer Thomas Telford. The shambles, or shemmels as we called it, was the local butchers' slaughterhouse. The new building at Merryton had a handsome frontage which was demolished along with the rest of the building in the 1970s to make way for a joiner's business.

Lands disposed of at this time were 42 acres at Moor Crook for £500, Whinnieknowe for £200 and ground adjoining the Shambles for £80. Three acres known as the Hempholes between Viewfield and the Links, as well as an acre of the Links, were also sold, and twenty lots of stances for houses on the Links were feued.

When, 160 years later, I pressed the District Council to purchase and restore the Town and County building, I was desperate to preserve for Nairn and its people a place that had been the centre of its civic affairs for all those long years on, or at least very near, a site where tolbooths had stood for centuries. I wanted to renew the life and the use of the building. I wanted our past to have a future, our people to retrieve their heritage, for, unbelievable as it may seem, the Town and County Building was until 1978 in the ownership of the Department of the Environment, this building having been erected in 1818 by the Burgh and County at their first and joint expense. It seems so appropriate that again, when Burgh and County are united as a District, it should come back to its rightful owners.

The present-day Council does not have such a struggle to defray expenses as the Nairn Town Council of 1818. They were dependent on the toll of the quarry, the toll of the parks, the toll of markets, the toll of the stable above the Shambles and so on, to make their people's lives tolerable and to give some form of elementary education to some of the children.

Nowadays, Council Tax and Support grants from the Government ensure that a reasonable standard of life is available to everybody, but one admires the spirit and foresight Councillors showed in these hard times when money was so difficult to come by and the Common Good Lands had to be put up for auction. They had ambitions for their town and went right ahead to achieve them.

3

THE FISHERTOWN

The River Nairn, from its source in Strathnairn in Inverness-shire, weaves down past the Clava Cairns and Kilravock Castle near the battlefield of Culloden, through the centre of Nairn to the harbour where it enters the sea. Although narrow, it is a fast-flowing river which in times of flood produces a spate that brings down whole tree trunks from the hills. It also brings copious quantities of fine sand which silt up the harbour basin and entrance, often making it difficult for boats to leave and enter. Many battles were fought by fishermen when they were operating the larger steam drifters, to get dredging done. There were times when boats had to ride at anchor in the firth until a high tide would carry them in, and sometimes they had to berth at Inverness and wait until it was possible to sail into Nairn.

When I was growing up, in the 1920s and 1930s, Nairn was a small but busy town, its economy based on fishing, farming and tourism. The fishing trade, both white fishing operating from Nairn harbour, and herring fishing worked around the coasts of Britain, contributed importantly to many other livelihoods in the town, although much of the gear and tackle for fishing was supplied by traders nation-wide. The High Street shops as well as other small outlets in the Fishertown were patronised by people on a daily basis and practised long opening hours. On the High Street at the time some shops, especially food shops, stayed open until 8pm.

The town had then a population of about 4,500. The rural area, the county of Nairn, had about the same number, as there were many farms with numerous farm servants and their families living in the countryside.

In summer the Fishertown was crowded with visitors. Many of the houses, or parts of them, were rented by families for self-catering, and in the case of people from Inverness they stayed for at least a month at a time. Some householders offered 'attendance', which meant cooking and cleaning for the visitors, and this attracted additional income. In time, because of the regularity of their coming, the visitors became an accepted and welcome part of the local scene and made friends all round the locality. They must have added quite a lot to the prosperity of the tradespeople over many years and Nairn High Street had thriving businesses. I remember thinking, in those years of depression, that shopkeepers were the only people who seemed to have money and I was very conscious of this among friends at school. The children of business people were always well dressed and seemed, in the austere atmosphere of the time, to want for nothing. Buying papers for Academy exams was no problem for them, whereas it was difficult for my mother to produce the twopence that we sometimes needed for them. These were books of blank paper for writing on. Each would be responsible for supplying our own.

The Fishertown had grown little by little after the building of the first harbour by Telford in 1821. From that time little fishing communities from nearby Delnies and Maviston moved to Nairn to benefit from the new facility and set up home here. The house where I was born and my mother was born, was built in 1821 and was then as far north as building in Nairn went. After that the space right up to the foreshore became occupied by the houses of fishing families. Most of the houses were built to the same proportions, being 40 feet long with a but and ben and two rooms upstairs with steeply sloping roofs. They were usually built in terraces to save space, and the end houses had their gables facing west or north against prevailing winds.

The new harbour attracted fishermen from nearby villages such as Delnies and Maviston, looking for better working conditions. Up until then, they had simply beached their boats on spits of land above high water mark. From that time, fishing as a local industry built up steadily, both in white fishing by lines, and as a base for herring drifters. Many boats were built for Nairn fishermen for these different purposes and a thriving community developed, all energies directed towards catching, curing and selling fish.

Nairn Fishertown lies below sea level and has suffered periodic flooding when very high tides and storms coincided. Once, early in the twentieth century, water spread almost up to the junction of Union Street and Society Street. Boys sailed their mothers' wash tubs in pools left by the tide along Firth Street and Park Street, just yards from the sandy bank of the beach where the sea had broken through. The bank was a poor defence for their houses, but due to shortage of building land in the neighbourhood in the late nineteenth century, and the need to be near their work, the people were willing to take a chance and built ever closer to the sea.

Each house in the Fishertown had a smoking shed with an earthen floor. Haddocks were strung by one lug on to a spit, and herring, because they were softer, were each hung by two hooks on a tinter (a wooden spar with hooks). When the fish were in place, fires were lit and kept going with durkins (fir cones). Smoking lasted for one and a half to two hours. In our young days, we used to collect bags of durkins from Granny Barbour's wood and sell them to the fishwives for sixpence or ninepence a bag. Fishermen or their older children dug for lug worms at the beach, to use as bait. Mussels were also used and they were gathered from groynes as far away as Munlochy Bay or Invergordon. These could be kept viable for a few days by being placed in a scaup, which was a little pool of water near the river, covered with a net and weighed down with stones.

After smoking, the fish were carefully packed into creels ready for selling. They were sold by size, for perhaps sixpence each. Very little profit was seen after bus or train fares were paid. Bartering sometimes took place and our neighbour Mrs Hastings used to bring home fresh butter or crowdie which she got in exchange for fish. For her rounds, the fishwife wore a flannel blouse, cardigan, a skirt down to the calf with a striped petticoat beneath, and black knitted stockings with black lacing boots or shoes. A black or dark tartan shawl was worn round the shoulders and old stocking legs were put over the jumper sleeves to keep them clean. All this work and travelling were done as well as housework and bringing up the children.

When line fishing came to an end in about 1924, some of the women bought fish from Lossiemouth and carried on their trade. Widows' pensions of ten shillings a week started in 1927 and they felt great relief at this as it took the pressure off them in old age.

In the children's world of the 1920s and 1930s Nairn was divided into two – the Fishertown and Uptown. I suppose that this divide, which was geographical as well as psychological, had its origins in some old animosity, but in our eyes at any rate the mock battles that we fought were largely idealistic. The physical dividing line between the Fishertown and the Uptown was the crossing of King Street and Union Street, and we used to set off with our weapons (usually sticks of some kind) to meet the challenge of our 'enemies'. The girls were easily intimidated and ran home at the first signs of aggression, but the boys brazened it out. No one, to my knowledge, was ever hurt, and the next day we were all off to school together, the best of friends. We were all poor, so we had that very important and binding condition in common.

The Fishertown was not only a distinctive community in which we lived; we were, of course, immersed in the activities of the industry from which it took its name, and we were also heirs to its past and its lore. Our parents and other relatives had a wealth of experiences and memories to recount which in a sense became part of our own.

At the height of its prosperity, early in the twentieth century, the fishing industry provided work and support for as many as 250 Nairn men and their 1500 or so dependants. Before World War I, 75 locally owned boats were engaged in either line or drift-net fishing, and in 1920, when the European market for salt herring was shrinking fast, the Mariner's Almanac for that year showed there were still 30 steam drifters and 42 fishing boats powered by sail belonging to Nairn fishermen. Even in 1931 there were still 210 men employed in the industry, notwithstanding a degree of emigration. By 1951 the census enumerated only 80 fishermen and today there is but a handful, none of them based in the town.

The fisher folk had a distinctive way of life, being to some extent detached from the rest of the townspeople by the nature of their exacting trade. They lived like a large family, observing a code of behaviour and set of customs and values prescribed by their seagoing forebears and handed down through generations. Their traditions were nurtured and sustained by a united and unswerving devotion to the non-conformist churches of their Christian faith, and by family and community loyalty to the ceaseless demands of the fishing industry. A stable pattern of life was established through close working partnerships and strong family ties, as boats were operated by groups of relatives who spent all their working lives together. The women shared equally in this solidarity in their closely packed Fishertown houses, communicating daily with each other over the men's work and their own connected duties. English was the language of school and church, Scots the language of home and work. The latter was a binding factor in their lives, because of the tools and utensils of domestic and work activities. As children we were familiar with and used much of the vocabulary, for example, brander (gridiron), bullace (axe), cruisie (oil lamp with rush wick), skaillet (shallow scoop with wooden handle), weers (wires, steel knitting needles), worsit (wool), kist (chest), caff (chaff), bush (the ropes to which drift nets are attached), bow (a fishermen's buoy), tippen (the twisted hairs that bind a hook to a fishing line), tinter (wooden spar with hooks attached for kippering herring).

The Fishermen's Society, founded in the eighteenth century and supported by contributions from members of the fishing community, which survived into our childhood, looked after the interests of families in distress after drownings at sea and loss of boats and gear. Money was lent for house-building, and the Society owned a mortcloth which could be hired for covering a coffin on its way to the cemetery. Before the hearse became fashionable eight men, in relays, carried the coffin all the way to the cemetery at the riverside. Charges for hiring the mortcloth varied between sixpence for a child, and two shillings for an older person.

Harshly self-disciplined, but kindly and generously disposed towards others, the fisher people preserved their way of life through good times and bad, supplying food and comfort to friends and neighbours in need, as naturally as to their families. It was common practice in my childhood to leave a box or part-box of fish on the quay for anyone who wanted to help themselves. In the 1920s, when I was growing up, depression and poverty set in due to the loss of the Russian and German markets for salt herring, and the tightly knit nature of the community began to weaken, as younger members moved away to other towns and other countries looking for work. Two of my uncles and their families emigrated to New Zealand. At the time it must have seemed as if they would never see their families or home again. Many friends went to Nairn Station to see them off, when they departed to the sound of the hymn 'God be with you till we meet again'. Granny died in her 80s, in an advanced state of memory loss, and never having seen again her sons and their families who had emigrated to New Zealand.

After World War II some members of our family were able to visit Nairn, and the attachment which they felt for home was passed on to the later generation who are keen to visit us. Granny kept a soup tureen on her dresser with the letters her sons wrote from New Zealand. These were very interesting as they contained much information about life and social conditions in New Zealand at that time. My uncle Sandy knew Peter Fraser, Prime Minister in the 1930s, who came from Fearn in the Black Isle. Mr Fraser used to call on our relatives when he was in this country, to pass on news of the family. When we were keen on stamp collecting, Johnny and I took Granny's letters from the tureen and tore off the stamps, putting the letters into the bin, so destroying what were valuable records of their lives in New Zealand. I have visited Island Bay near Wellington on the rocky coast where they settled. There was no shelter for boats there and my uncles' boat, uninsured, was swept away and wrecked in a gale. This was a serious misfortune, taking away their means of livelihood, and led to poverty even harsher than that which we experienced on the collapse of the fishing industry in Nairn. One of my cousins told me that sometimes they had no money or food in the house, and survived only because they were given credit at a small corner shop owned by Italians.

4

THE DAYS OF THE
WIND-POWERED FISHING BOATS

Although only a few small fishing boats relying on sail power still operated out of Nairn in my youth, the days of sail were still very real to my parents. My mother in particular spoke of them with real regret rather than of romance. The safety of the crew, often all of the male members of the family, was a constant concern. She often recounted how Granny would go to the beach in the middle of the night to look for signs of the family boat nearing home in stormy weather.

Most families who owned a fishing vessel from the 1880s up to the outbreak of World War I prospered. A rail strike in 1891 and a fall in the value of the Russian rouble at about the same time were minor disturbances, which did not affect the secure and financially rewarding herring trade. Before the days of steam, Nairn fishermen went far from home to the fishing grounds in open-decked fishing boats with sails, the famous Zulus. These were built in various sizes to suit the different types of fishing, line or drift net, and were always painted black. The Zulu boat came into use just about a hundred years ago and was greatly liked and admired by fishermen in the North-East.

It got its name from the Zulu wars, which were raging in Africa at the time. It was a highly manoeuvrable boat with a big area of sail and large hold for fish. The big Zulu used for the herring fishing was 70 feet long with a beam of 20 feet and usually carried a crew of six men and a cook who all lived in the cabin.

It had a coal-burning stove for cooking, and bunks for men to sleep in, and a large kist holding stores was screwed to the floor for use as a table, while men sat on lockers around the kist to eat. Ballast stones for the Zulus were gathered from the beach, but as salt water got in on them and they caused a smell, they were removed after each long trip, hosed down with water from the boiler, and left out to be freshened up by rain. Zulus were fast and have been known to sail from Great Yarmouth in Norfolk to Nairn in 48 hours.

My mother told me that her brother Isaac went as cook at the age of about thirteen, on her uncle's boat to Yarmouth. The skipper, also Isaac, was a demanding boss, always harassing the crew at their work. He had given young Isaac an awful row, and he, always rather fiery-tempered, took a pailful of water and threw it over him. When they got back from the fishing, Granny went to Uncle Isaac and gave him a severe dressing-down for treating her boy so badly. Boys served apprenticeships as cooks, before they became deck hands and part-owners of their own boats.

There had been a long tradition of bringing presents home from Yarmouth, the port of the autumn fishing. The shops there were well patronised by the Scots fishermen and girls who went for the gutting. In contrast to the small and rather limited selection of goods at that time available in the small north-east towns of Scotland, the large department stores in Yarmouth, with their carpeted floors and stairs and great variety of merchandise, thrilled and excited the women. With money to spend and bottom drawers to fill, they spent freely and came home with tea and dinner sets, dresses and button boots, ornaments and table covers.

The men brought presents for their families. Boys got cricket bats, footballs and football boots, torches and games. Girls got dolls, watches, gloves and purses.

Fishing boats relying on wind power were cheap to run, so steam was used only for winding ropes and hysing (hoisting) the foremast and sail. Zulus were built at the east-coast ports of Hopeman, Lossiemouth, Buckie, Banff, Fraserburgh and Macduff. Most fishing ports had their own shipbuilding yards at that time, and although Inverness also had yards, no herring boats with sails were built there. All Nairn fishing boats, however, were registered at Inverness, bearing the letters INS followed by a number.

5

THE STEAM DRIFTERS

The first steam drifter to come to Nairn was *The Hope* in 1904, making the change from sail to steam, and this change-over continued until the whole fleet was steam-operated. A lot of shipbuilding went on at the Rose Street Foundry, Inverness, especially of drifters. These were wooden vessels with steam engines which were extremely reliable, hardly ever needing repair.

Although the Zulu boats had worked fairly efficiently, their performance was related to wind and sail. The steam drifter had a motive power far superior to sail and provided much better living and working conditions for the fishermen. Independent of wind, she could go and come at will, staying in port only in the very worst of weather.

The average herring drifter was between 80 and 86 feet long, with a beam of 20 feet and a draught of 10 to 12 feet. She was of standard design with certain internal variations to suit the wishes of individual owners, who wisely never tampered with the shape of the hull, which had a fairly straight stem. She carried a lifeboat and main mast which, along with the derrick, rested in fitments on the roof of the wheelhouse. The mizzen sail, aft, was always raised when hauling. The fo'c's'le was built as accommodation for crew and two or three men slept in it, but although suitable enough in calm seas, it was uncomfortable in rough weather. In the drifter *Glenerne* four men slept in the fo'c's'le and six in the cabin, but in *The Brighton of the North*, a bigger and more up-to-date vessel, there was accommodation aft for all ten of the crew.

> The best time to catch herring is when the sun goes down. The darker the night the better.
>
> Sandy Baillie

The use of the steam drifters made no difference to the catches of herring. Fishermen at this time were working 'blind'. Echo sounders did not come into general use until the 1930s and until then fishermen depended for success upon instinct, observation and knowledge of the fishing grounds. If the water looked oily, this was an indication of a midwater or surface shoal, and the presence of diving birds also suggested there were herring about.

The herring industry had passed its peak when my parents got married in 1921, and the trade settled in to a time of severe depression. By 1937 it had all come to end as far as herring drifters were concerned. There were no markets for their catches and the boats were out-of-date and in poor repair. Many of the fishermen in the North-East decided to change over to seine-net fishing and acquired motor-driven vessels for this purpose.

In 1977, some friends of the Fishertown set up a museum telling the story of the community's past. In addition to conserving the history of the fishing and providing a source of reference for present and future generations, the museum helps explain why the Fishertown is so different in character from the rest of the town. It is sad that reasons for its being different no longer exist and that fishing can now be seen only in a museum setting.

6

THE SEAMAN'S HALL

The centre of Nairn fisher people's social and cultural life in the later nineteenth and twentieth centuries was the Seaman's Victoria Hall in Harbour Street. After many years of careful planning and fundraising, and with the help of prominent townspeople and friends from the landward area of Nairnshire, it was built and opened as a temperance hall in 1887, Queen Victoria's Jubilee year. Most fisher weddings were celebrated there, the religious ceremony followed by a reception and dance, and these were occasions of great joy and happiness, lightening the often grim and anxious years of worry and hardship.

The Hall was run by a committee, which in the early days included the Provost of Nairn, a Baillie and the Editor of the *Nairnshire Telegraph*, Mr George Bain. Mr Bain drew up the constitution. The Hall is still managed by an elected committee with a President, Secretary and Treasurer.

The Seaman's Hall was also the main meeting place in Nairn for the local Temperance Society who drew many enthusiastic children to their 'Band of Hope', where leaders told them stories and taught them temperance songs. The children signed the pledge and wore the blue ribbon, token of their vow to abstain from strong drink, and many kept the promise all their lives. On New Year's Day each year, before World War I, a Blue-Ribbon march was held. Led by a piper, with the Temperance flag flying, the marchers, starting from the Hall, made their way around the streets of Nairn. In my childhood these activities had largely ceased, but the practice of temperance was as strong as ever.

In the late nineteenth century, the Reverend Mr Burns, then minister of the Parish Church, conducted the choir for a time. He required auditions before accepting members and tested women's voices while they were redding and baiting their lines. Sadly, before my youth, the choir had ceased to exist, doubtless partly the result of the demoralisation which had hit the community, but also perhaps because of the prolonged and frequent absences of the men at the herring fishing and the accompanying migration of the young women for gutting and packing the herring.

There was the Seaman's Hall Choir, able and ambitious enough to render, on at least one occasion, the Hallelujah Chorus from Handel's *Messiah,* learned and sung from tonic solfa. My uncle, Hugh Wilson, conducted the choir at the age of only twenty-one. He and his brothers, who were enthusiastic singers, understood and used a modulator to train the choir members. They were good customers of the music publishers Messrs Bailey and Ferguson in Glasgow, who supplied all the sheet music they required.

The modulator was a chart used in the tonic solfa system for exercise in sight-singing. It looked something like this and was still being used by singing teachers when I was in school:

```
doh
te
lah
soh
fah
me
ray
doh
```

Salt Herring on Saturday

The teacher or conductor pointed to the notes and the singers were expected to sing them in any order as he pointed. This was also an exercise in pitch at which people of the Fishing Community became expert, contributing valuably as members of the local church choirs.

Hugh Wilson, who, through his lessons on navigation before World War I, had prepared some dozens of young men to steer their boats safely through seas all around Britain seeking out shoals of herring, had also, unwittingly, prepared them for four years of naval warfare in distant and dangerous waters. He became a commissioned officer in the Royal Naval Reserve, serving in the Mediterranean, and was later awarded the Distinguished Service Cross. He returned from the Dardanelles on leave, to be married in the Seaman's Hall in 1917. The *Nairnshire Telegraph* reported that the ceremony was a bright and happy one, and the ninety guests were given a sumptuous dinner in the Hall, after which a dance was held.

The Hall had a quite extensive library from which men could borrow a box of books for one shilling to take away on fishing trips. A lot of reading was done, particularly at weekends, on board the boats. The books were very largely of the 'improving' kind, the novels being mainly classics, and there was a wide range on political economy, history, philosophy, biography, politics and religion. The fishermen were well informed on the last two. They had a highly developed interest in politics and Nairn Fishertown was a stronghold of Liberalism while part of the old parliamentary constituency known as 'Inverness Burghs'. By the time that I was in my teens the library was little used and had not been added to for many years: another sign of decline in community cohesion.

Many efforts were made throughout the generations to

raise money for the upkeep of the Hall. Sales of work, whist drives, weddings, concerts and soirées, and lets to various organisations financed heating, lighting and maintenance. Grants from the Council were not available until late into the twentieth century. Tickets for membership of the Hall were sold around the Fishertown. I remember my father being unable to produce the three shillings required before the war.

Nairn fishermen, fully literate following the Education Act of 1872, and with easy access to a variety of books and periodicals in the Seaman's Hall library and reading room, took a lively interest in politics and eagerly attended meetings in the Hall, their thoughts and ideas on the subject sharpened by preliminary discussions and arguments when groups regularly gathered round a bench at the Harbour Master's Office to discuss the news. This gave rise to the nickname 'Harbour Parliament', and it remained in being until fishing had almost ceased and the area was redeveloped at the end of the twentieth century.

The fishermen were also strong churchmen, attending each Sunday when at home two services, bible class and, at Rosebank Church, a gathering of young people called 'Happy Sunday Afternoon'. The Seaman's Hall also accommodated Evangelists' meetings, and, towards the end of the nineteenth century, the Sabbath School of the United Presbyterian Church. An annual Spring Flower competition and show was also held, and so the Hall was a much needed and well used facility.

In the 1930s, Nairn Town Council erected groynes at regular intervals along the beach to assist the build-up of sand, and also strengthened the bank with rubble refuse to inhibit erosion. A strong growth and spread of marram grass then took a firm hold on the sand and provided a much better protection from the sea. The 'piles', the local name for the original wooden groynes, seemed purpose-made for the children, who used them as diving boards when the tide was in, and climbing frames when the tide was out. We spent many happy hours around the piles. Even in my young days half of the Fishertown was frequently under water as the river flooded through the ground water pipes. In spite of a huge new sewerage scheme, this problem has not been entirely eliminated, and the Park Street and Harbour Street areas still flood at times.

In some parts of the Fishertown there is an attractive and unusual arrangement of houses, where pairs of gables give on to the main streets, a feature much admired by some architects. The houses are built very close together with little spare ground between them, for the trade did not require land as much as it did buildings – sheds for smoking fish and sheds for storing gear. Some activities, such as baiting and redding lines, were pursued outside the front doors. If one dug the soil outside many of the houses the remains of mussels, a common bait, could still be found today.

The house I live in was just like this at one time with a smoking shed attached and a large shed opposite. The smoking shed was knocked down in 1946 and a kitchen and bathroom built in its place. The house was erected in the early 1820s, probably in response to the new harbour, and the big shed about 1899, when the herring industry was at its height and extra space was needed to store the nets and ropes used by my grandfather and his sons.

I saw my brother Johnny soon after he was born, being held upside down by the midwife, Mrs Main, a widow, who lived just up the road from us and who, incidentally, also 'laid out' people after they had died. After that I watched him being bathed. Babies were always born at home in those days. A maternity ward was opened in Nairn Town and County Hospital in 1932 and most births took place there from that time until the 1980s when the ward was closed and maternity needs were met by Raigmore Hospital in Inverness.

A few weeks later I was present with the rest of my family at Johnny's christening in the Rosebank United Free Church. Dressed in a sailor suit with 'HMS Monarch' on the hat band, I was taken to the church that day by an aunt and uncle. I was three and a half years old. The minister, after the christening, endorsed the birth certificate with these words, 'The child bearing the name on the other side of this paper, has been baptised by me according to the rites of the Presbyterian Church'.

Johnny was far more to me than a doll, and I remember only one doll I had, much later, with a china face. I had just been given it when one of my brothers took it from me and dropped it on the lino-covered floor. That was the end of it and I expect I displayed plenty of anger at the time and never played with dolls again.

7

THE CHURCH

Johnny was born and christened in 1928. By 1929 we had left the Rosebank Church and were now members of the United Free Church (Continuing). A move to unite the existing UF Church with the Established Church of Scotland caused splits in congregations throughout Scotland. A faction in the UF Church opposed becoming part of the Established Church and most of our family in Nairn were part of that dissent. As in other towns, the trauma of the 'Union' (for which the majority of the Rosebank Church voted) divided families, neighbours and friends, and resulted in much bitterness which took many years to evaporate; but the rift was probably more damaging in our community than in most places.

The Church, in addition to its primary role in the social life of the community, cemented the solidarity which was an essential part in the hazardous occupation that was fishing. Fisher people were deeply involved in the running of their churches and provided a large part of their leadership. The 'Union' ended in division of families, neighbours and friends. There was bitterness in the fishing community and this lingered for many years. Men and women, who had lived and worked in close association for most of their lives, were now ill at ease with each other. I felt this sense of estrangement throughout my growing years. A group which had hitherto been members of a self-confident majority was now a homeless minority.

My father was convinced that before the vote for the union, the dissenters were in the majority, but that the minister, the Rev William King, had persuaded enough of them to change their minds so as to secure a vote in favour of establishment and retention of the church building. No doubt there was more security for the clergy in this decision. The animosity then engendered has now died out completely because, of course, few people remain who remember the cause of it. All of my mother's family, and most of my father's, left Rosebank and, along with a good number of other dissenters, set up their own continuing church in the Public Hall. The church building went to the majority who remained at Rosebank.

A very happy and successful church was set up in the Supper room of the Public Hall on Nairn High Street. Among the congregation were a number of Nairn businessmen and various retired and moderately wealthy men and women. They were generous in giving. At a time of severe economic depression the rest of the congregation, fishermen and tradesmen, had little money to spare and many expenses had to be covered. They called a minister, Rev Peter Mackintosh, who had a wife and three children. His stipend and all the expenses of running the church had to be met by the congregation. But there is no doubt that it worked well. Mr and Mrs Willie MacGillivray generously gave the use of their house 'Redesdale' on Waverley Road as the UF Manse.

I used to be sent off early for the Sunday morning service with Granny and we sat very near the front where I had a good view of Mr Alex Napier facing the congregation to lead the praise. He had a neat pointed beard and a waxed moustache with twirly ends. He and his family were very musical. He had two daughters, Ella and Mary. Ella was an accomplished singer and Mary played the organ. Mary taught me in the Millbank Primary School and when I was in her class I remember having to memorise Psalm 46, 'God is my refuge and my strength'. I got a threepenny bit for being able to recite it the same day. Mr Napier ran the Nairn Ladies Choir which, before World War II, broadcast from Aberdeen on the radio. This was a great honour and we all listened in. I was thirteen at the time and was thrilled when Mr Napier asked me to join the choir. I think he must have died soon after because I never did sing with them under his leadership.

Metrical Psalms and Paraphrases (passages of scripture adapted and set to music) began and ended our services every Sunday. We sang from a Psalter which contained the words of the Psalms and Paraphrases and many tunes to which they could be fitted. The pages of the Psalter were split between the music and the words, so that they could be conveniently matched up to the required setting. For instance, there were at least four tunes for the psalm 'The Lord is my shepherd' – *Crimond, Wiltshire, Orlington and Martyrdom*. We often sang a hymn I disliked very much, but it seemed to be a favourite with the Minister. The opening lines were 'Childhood's years are passing o'er us, soon our schooldays will be done'. The words seemed to constitute a threat to our carefree, happy days which we did not wish to end. My mother felt the same about 'The sands of time are sinking, the dawn of heaven breaks'. It was as if life was over before it had properly begun.

On Communion days, the elders were dressed in black morning clothes to serve the bread and wine. The Common cup was used at that time. When we were in our teens we sometimes attended on Communion day, but sat at the back of the congregation just watching. My elder brother Alex attended church twice every Sunday and also Sunday school, spending at least three hours weekly in church. I never heard him complain about it or express reluctance to go. Johnny and I were left at home with Granny whilst the others went to church in the evening. We used to play rough-and-tumble games in her room until she was provoked on one occasion to cry out, 'The Lord'll send a curse on you!'. She would look after us only to let my mother out to church or a church meeting or other religious activity.

Highlights of the church year for the children were the Sunday School parties and picnics. Many of the adults accompanied us to the picnics each summer and organised the games – cricket, football, rounders and the races. We used to be dressed hours beforehand waiting impatiently for the time of departure to come. A favourite place for our picnics was Darnaway where we had the use of the field next to the village hall, very handy when it rained. Everything was free: transport, cakes, lemonade, tea and ice-cream. It made a very welcome day out for harassed mothers. Other favourite picnic spots for us were the Culbin Sands, Brodie Castle and Fochabers. We frolicked amongst the dunes at Culbin. Afforestation had not progressed very far in the '30s and there was nothing to be seen around us but sandhills and bents. We were surprised to find tadpoles thriving in little pools of fresh water. This was about three miles from the sea and it was easy for us to believe that this was the 'little Sahara desert'.

Our Christmas parties were held in the Free Church Hall on King Street. We usually rehearsed little plays and songs/ hymns with our Sunday School teachers and dressed up to perform. We were each given a paper bag containing sand-

wiches and cakes, and always received presents from the
Christmas tree which was lit with real candles. Again many
adults attended and enjoyed the parties. These were splendid
occasions for us because of the lack of social activities during
the rest of the year. One Christmas my elder brother Alex was
in bed with scarlet fever. Johnny and I grumbled and complained the whole day long because we were not allowed to
go to the party and risk infecting other children. A year was a
long time to wait for the next one. However, our Sunday
School teacher delivered our presents that night.

It had not yet become the fashion to have Christmas trees
in the house. Our first one was a fir tree branch our neighbour
Mrs Hastings brought from the countryside. I can't remember
how we decorated it, but to us it smelt of Christmas and was
the start of more colourful celebrations in the future. My
mother always did a big clean on Christmas Eve. The
chimney was swept 'so that Santa wouldn't get too dirty
on the way down', and then the whole room was scrubbed
out. We hung our stockings up at the high mantelpiece and
left a glass of ginger wine and some cake on a stool for Santa.
The men were always at home from the fishing for Christmas
and my father used to sit by the fire and call up the chimney to
Santa to tell him what we wanted. This was very convenient
because he relayed to us answers which complied with what
we were likely to get rather than what we most desired. There
was always a bought net Christmas stocking for each of us,
filled with cheap little items which became more and more
worthless as the day went on.

There was sometimes a chicken for Christmas dinner, or a
steak pie and another occasional treat, Bartlett tinned pears
with custard. The chicken was always boiled so that soup
could be made from the stock. At that time chickens were

expensive and we only had one very occasionally. Christmas was a very low-key affair in Scotland in those days. Tradespeople worked on Christmas Day and the shops were open. As far as I remember there were not even services in the Church of Scotland or UF on that day. Certainly there wasn't much fuss made of Christmas in our house during our childhood. We had not yet broken away from the restraints of the Free Church tradition, and I don't remember Granny acknowledging Christmas at all.

As my family were all teetotal, there were no New Year celebrations in our house. New Year was associated with drink, so was pretty well ignored. Later in my childhood, I remember being shocked to see a drunk man lying on the road near us one New Year's morning. We were none the worse for disdaining New Year celebrations. We really didn't know we were missing anything.

The Moderator of the UF Church once came to visit the congregation. The Rev Mr Goldie wore a very fancy tasselled gown. This inspired me to write a letter to our Sunday School magazine, *Cheerio*, and I was very pleased to see it in print in the next issue. I was nine or ten at the time. At the end of the Sunday School sessions, Mr Ingram, a member of our church, always presented to each boy and girl of the Sunday School a large block (probably half a pound) of plain chocolate. I think it was called *York Motoring Chocolate* and it was much more bitter than modern plain chocolate. We hated it, so much to our chagrin, each year we grumpily put it aside, to be eaten probably by our parents. If it had only been milk chocolate, how welcome it would have been! Mr and Mrs Ingram were English; he had been a butler and she a ladies' maid in some big house. Retired, they lived in Lily Cottage on the Inverness Road. We loved being invited there – keen to see how they lived. They were very hospitable. Mrs Ingram wore high-necked dresses, with her hair bunched up on top of her head; her husband was always very perjink (precise, particular), as would be expected of a butler. They used to give my mother apples from their garden, and at least once I remember Mrs Ingram gave her money to buy sugar for the jelly making and, needless to say, this was a very welcome gift.

Another highlight of the church year was the adults' 'social'. This was at the AGM of the church, and after the business was over, they enjoyed visiting speakers, music and a delicious tea. My parents looked forward to this annual event. They always contrived to bring home a special cake for each of us. Later on, as a member of the church choir, I enjoyed these meetings in my own right. With the commencement of World War II in 1939, we had to move out of the Public Hall as it was requisitioned for military use. We went to the Welcome Hall, owned by the YMCA, in Harbour Street. What I remember most about the Welcome Hall was its dark green furniture. It was a small hall and our congregation fitted nicely into it. There we stayed until invited at the end of the war to join for worship with the Congregational Church. Our minister also left to take up a charge in Edinburgh.

8

HOUSE AND HOME

My two brothers and I were born in the 1920s when the fishing industry was in severe decline. Steam drifters were getting old and demand for herring was poor. Overseas markets had been lost because of the war and the Ottawa Agreement of 1932 on imperial preference which meant that countries of the British Empire kept import rates lower for each other. It put home producers first, Empire producers second, and foreign producers last. This continued until replaced by the General Agreement on Tariffs and Trade (GATT) after World War II.

My parents married in 1921. My father had brought home from Malta (where he had been based as part of his service in the Royal Naval Volunteer Reserve in World War I) a length of pale grey silk for my mother's wedding dress. The dress had been trimmed with tiny pink silk roses. Years later as children, rummaging through bags and boxes, we found fragments of the dress which had frayed and fallen into pieces. We extracted some of the rose trimming and, showing it to mother, heard the story of the dress. She also had a beautiful Maltese silk bed cover which has remained in good condition ever since. My father and mother married in the aftermath of World War I when housing was in short supply. They had spoken for a house at the Links, but it was withdrawn from sale. My grandparents gave them the use of rooms at 27 Union Street, but in the end this became their permanent home. My grandfather died before I was born, and then, I suppose, it was expedient for the family to stay on as company for Granny.

There was cold running water in the house, but no inside toilet. This was in the big shed opposite, built at the end of the nineteenth century mainly to accommodate fishing gear. The house dated from 1821. Going to the toilet on dark winter nights was awkward, as we had to be accompanied, lighting our way with a torch. In winter my mother lit a small paraffin lamp to keep the toilet free from frost and this worked very well. The cistern and bowl were sensibly made from cast iron. Later on, they were replaced with ceramic fittings, which shattered with the frost and made the place unusable. We have long since had indoor facilities. We used to cut up old newspapers into small squares and string them up for use in the outside toilet, but sometimes had rolls of Izal toilet paper.

Washday was Monday because we always had a clean 'shift' of clothes at the weekend. All the washing was done in the shed where there was a big open fireplace. A huge enamel pan of water was heated on the fire and emptied by a skillet into the wooden wash tub. Fairy and Lifebuoy (carbolic) soap were used to scrub on the wooden scrubbing board and, after rinsing, the clothes were mangled. We had a very large and heavy mangle with wooden rollers. When we were on holiday from school, my mother was glad to have our help with the mangling and the thumping of blankets in a wooden barrel with a wooden dolly (many of the washing utensils were wooden and very heavy). She started washing when we left for school in the morning and was still at it when we came home for dinner. She would continue in the afternoon. Quite apart from the effort of hand washing, the handling of the heavy wooden barrel and tub required considerable strength and was energy-sapping. I remember my mother falling asleep in the evening of a washday. We had long washing lines and a drying green where the output of the washtub was hung out to dry. I remember my mother's despair when it rained, as all the results of her day's work got wet again and there was so little room inside to dry things. Sheets, pillowcases, towels and all our underclothing, so painstakingly laundered, had to take turns on the wooden clothes horse near the fire in the living room to be gradually dried off.

Mending took up a lot of time too. Nothing was put away until holes were patched or darned. I well remember the smell of washing being dried indoors and the herring basket full of articles waiting to be ironed. Ironing facilities were primitive. Heavy irons had to be heated at the fire. There were always two irons so that one could be heated while the other was in use. The kitchen table was the ironing board, with a thick layer of old flannelette sheet as padding.

Before being folded away, everything was aired on the clothes horse. Not long before World War II my mother acquired a gas iron with a flame inside heating the plate. This was a big improvement and speeded up the whole process. Soon after we had electricity installed in 1945, the electric iron took over. Later a small Hoover washing machine with hand wringer was acquired and this took much of the labour out of wash day. In the summer when the beds were all stripped down and all the heavy bedclothes washed, we carried the blankets in a herring basket to the Links and spread them on the whins, just about where the children's playground is now. They were left to dry in the sun. I don't remember any being stolen. Mattresses, which were made from horsehair or flock, were removed from the beds, aired outside and beaten to remove the dust.

The condition of the doorstep was an indication of pride in one's house; in the Fishertown many front doors (in most cases the only door) opened directly onto the street or close and the doorstep was an important feature. My mother kept a block of 'freestone' for treating the doorstep after it had been washed; rubbing the surface with this 'stone' produced a light uniform finish which made the house seem cared for; unfortunately the first soiled foot stepping on it spoiled the effect completely.

We slept with a bottom sheet and then the blankets over us. Whether this was because sheets were scarce or whether it was for warmth, I don't know. In the wintertime my mother filled stone hot-water bottles and placed them between the blankets and our feet. This way they stayed warm most of the night. Our bedroom was unheated and I remember putting my head down on ice-cold pillows. We slept under the sloping roof of the house where it was very hot in summer and very cold in winter. One night going up to bed with a candle, Johnny and I set fire to the wallpaper on the landing. It was quickly put out, but we were given a very severe warning for the future. I still use the bedtime prayer we were taught as infants:

> This night as I lie down to sleep
> I pray the Lord my soul to keep
> If I should die before I wake
> I pray the Lord my soul to take

After asking God to bless all the members of the family and our friends, the prayer concluded thus:

House and Home

> Lord keep us safe this night
> Secure from all our fears
> May angels guard us while we sleep
> Till morning light appears. Amen

Our first taste of live theatre was from the Bertha Waddell Children's Theatre. This was a group of players based in Glasgow who went around the country presenting dramatised fairy tales, songs and poems. They performed in the Public Hall in Nairn. I saw them once as a child and still remember the magic plant episode, when the players contrived to show us a plant growing as we watched. This was real magic for us, but probably very easily manipulated by the actors, who were all dressed in colourful costumes, with hand-drawn carts that all seemed straight out of our storybooks. Much later, when I was teaching in a Glasgow school, I went with my class to a Bertha Waddell concert. It was still lovely to watch.

The first film that I saw was *David Copperfield*, in black and white, at the Playhouse cinema in Nairn. My mother took me and I was thoroughly won over by the performance. Freddie Bartholomew played David; Mr Murdstone was really frightening and it all made a great impression on me. When we came out of the cinema, there were six inches of snow on the ground, which made the whole experience unforgettable.

When the Regal cinema opened before World War II, its first film showing was of *A Tale of Two Cities*. Johnny and I, with enough money for the cheapest seats, joined the long queue. The unusually large crowd pressed in on us and only a few people were allowed in at a time when the door was opened. We were pushed up against the door and I was afraid that Johnny would be crushed. I appealed to the doorkeeper to let us through and we walked into the cinema vestibule where we were transfixed by an inviting blue glow of light from upstairs, but there were no seats for us. We did get in, I think, next day, and were magically transported to revolutionary France by Ronald Colman playing Sydney Carton.

The other memory I have of the cinema from my childhood is of a visit from school to see some historical presentation about Queen Elizabeth I. What I remember best about this was the musical theme running through it which sparked off a lifetime's pleasure and appreciation of classical music. It was Purcell's 'Trumpet Voluntary'. Every time I hear it I go back in time to that film and the wonderfully inspiring tune which woke me up to the joy of orchestral music.

1. Nairn fisher girls at Great Yarmouth, 1923-4. Second left: Jessie Main; third left: Annie Main; fourth left: Janet Milne.

2. A fishing lass tying bandages round her fingers in readiness for herring gutting, to prevent cuts from the gutting knife. Drawing by Sonia Stevenson. Scottish Life Archive, C7254.

3. Herring gutters at Stornoway, working one of the old-style 'farlins',
c. 1900-10. The little round buckets were called 'croosies'; they were later
replaced by tubs with handles. Scottish Life Archive, C7257.

4. A team of gutters outside their hut, 'Glenerne', at Gremista, Lerwick, Shetland, c. 1900. 'Glenerne' was the name of one of the fishing boats. See Fig. 8.

5. Five Nairn gutters, ready for work. Left to right: Frances Coggs, Maggie Jane Barron, Jessie Downie, Annie Jamieson, Jess Ralph, c. 1920.

6. 'Sammy Jo' Jamieson carrying two 'skus' (sculls) holding lines
baited for fishing, *c.* 1916.

7. Mr. and Mrs. Laird, and son Hugh, with Willie Downie at
24 Union Street, Nairn. Hugh is 'redding' lines, 1922.

8. The family drifter, *Glenerne* INS 27, at Nairn in the 1930s.
Scottish Life Archive, C7258.

9. Mrs. Hastings, Union Street, Nairn, with smoked fish
ready for selling, *c.* 1936.

10. Women dealing out shares of fish at Nairn harbour, *c.* 1930. Their hurlies can be seen in the background. Left: Mrs. Isaac Bunker; centre: Mrs. Cope (Annie Mac); right: Mrs. A Barron. Photo lent by Mrs. Storm. Scottish Life Archive, C7702.

11. The Main sisters outside their thatched house. The smoking shed is on the right. Pre-World War I. Scottish Life Archive, C8705.

12. Fishwives with creels at Nairn harbour, *c.* 1930.
Mrs. Hastings; centre: Mrs. Finlayson; right: unknown.

13. Nana Duggie and Mrs. Cope, selling fish on Nairn High Street, 1930s–40s.

14. A shed opposite 27 Union Street, Nairn, for storing fishing gear.
Drawn by Elaine Dixon, 1981.

15. Fisher folks' houses in Society Street, Nairn, gable end on to the road.
Drawn by Elaine Dixon, 1981.

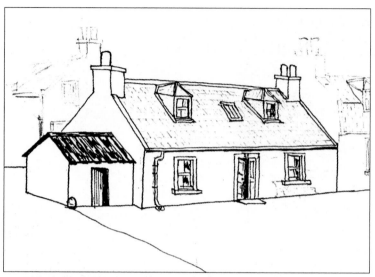

16. 27 Union Street, Nairn, pre-World War II, with smoking shed attached.
Drawn by Alexander M. Bochel in 1982, from memory.

17. 27 Union Street, Nairn, in 1953. A flat-roofed kitchen (left)
has replaced the smoking shed.

18. The 'maister' at the school. Mr. Richard Jamieson, with his pupils.
The author's mother, Margaret Wilson, is second left in the back row, *c.* 1900.
Scottish Life Archive, C19123.

19. James Asher, baker and confectioner, who supplied the food for
Fishertown weddings. Pre-World War I. Scottish Life Archive, C6140.

20. The writer's grandparents, Margaret and Hugh Wilson,
with two daughters and granddaughter, Margaret Ellen, *c.* 1918.

21. Granny (Margaret) Wilson, at the door of 27 Union Street, Nairn, *c.* 1930.

22. Granny Wilson with the family at Brighton House in 1926, just before the sons emigrated to New Zealand. The writer, Margaret Bochel, is on the cushion in the front row.

23. Margaret Bochel, aged 10, with Granny Wilson, at Brighton House, 1934.

24. Alex, Johnny and Margaret Bochel
with parents, at 27 Union Street,
Nairn, 1938.

25. Johnny Bochel with his bicycle
at 27 Union Street, 1943.

26. A fishing boat, INS 332, in Nairn harbour, dressed with flags for a wedding in the family, 1905. Scottish Life Archive, C5551.

27. The Seaman's Hall, Nairn, at the present day.

28. The 'contrack', or agreement to a wedding.

29. 'The 'biddin'', or invitation, to a wedding, late 19th – early 20th century.

30. Lieutenant Hugh Wilson and his bride,
at the Seaman's Hall, 1917.

31. Mr. And Mrs. David Wallace, Comrie Villa,
after their wedding in the Seaman's Hall, 1921.

32. Seaman's Hall choir, *c.* 1910. The conductor, centre front row, is Hugh Wilson.

33. Pledge of abstinence of the Temperance Society, in the Seaman's Hall, 1987.

9

GRANNY WILSON

My Granny Wilson was born about the middle of the nineteenth century, the eldest of a family of three boys and two girls. Like most fisher families at the time, their livelihood depended on line fishing. Groups of four men owned yawls from which they set lines each day in the Moray Firth. Much of the preparation for fishing and later the selling of fish were done by the women and older girls and boys. The whole house and family were employed to keep one man going to sea.

Granny did not speak much about her early life, but from what my mother told me it was assuredly one of unmitigated toil and penury, of walking miles every day with a heavy creel of fish to sell around the countryside, and coming home with a meagre return.

Religious observance was strong among the Nairn fisherfolk, and although Granny learned to read at school, but not to write, her education continued to progress into adulthood thanks mainly to the United Presbyterian Church. For relaxation and, no doubt, edification, she read *The Sunday Companion* and *The Christian Herald*. When I was a little girl, I used to accompany Granny to the United Free church service in the Public Hall. She dressed in a long black coat and a lovely black silk toque tied under her chin with broad black silk ribbons. She wore fine black-laced boots. I don't think she ever owned a pair of shoes.

Granny had brought up a family of six at 27 Union Street, although she had given birth to ten children. The first four were boys, the fifth a girl who died at just a year old after vaccination. Granny stopped my mother from having us vaccinated because of this and we had no smallpox vaccinations until we were grown up. Granny had another girl in 1885 and then, in 1888, triplets, two boys and a girl who all died within days of their births. I used to think they must have died of cold until I discovered they were born in July. They were premature and probably had breathing difficulties. My mother was born in 1890 and was the youngest in her family. With four sons and a husband at sea and two young girls at home, Granny had endless tasks to do every day. All work had to be finished on Saturday night so that Sunday was kept for worship and quietness.

As children, we were allowed no secular activities on a Sunday, no singing except hymns and psalms. Reading the *Children's Newspaper* or going for a walk was our only recreation but we enjoyed that because we met up with friends and explored the riverside and beach.

We had in the house an American organ, bought for my mother by her brothers. She had at one time had lessons on it. Granny's sons were musical, one especially gifted for he composed tunes and taught the whole family to read solfa notation from a modulator. When they were at home from sea, there was much singing in the house, to which, my mother told me, people outside stopped and listened. When my brother Alex became a teenager he played the organ a lot, teaching himself to read solfa and later staff notation when he wanted to play music more sophisticated than hymn and psalm tunes. Very often on Sunday afternoons various friends who liked to sing came to our house and we all sang Sankey's hymns around the organ. We were able, without difficulty, to find all the parts, soprano, alto, tenor and bass. Johnny remembers lying in bed at night while the grown-ups practised quartets to be sung at Congregational socials. I started early singing the alto line, having been taught to read solfa seated on my father's knee. He used the psalter, and a tune I well remember was St Asaph, which occupied only one octave and was quite tuneful. I would have been six or seven at the time. The words to it were paraphrase 66:

> How bright these glorious spirits shine!
> Whence all their white array?
> How came they to the blissful seats
> Of everlasting day?

This paraphrase was sung at times of mourning such as Armistice Day and at Easter time. When my father was dying, he told me that he passed the time fitting the psalms and paraphrases to their tunes.

This immersion from a tender age well into my teens and twenties in solfa notation resulted in a useless ability to turn any tune into solfa notes. This still happens all the time and sometimes spoils my enjoyment of music. Still, this method of reading music, if properly understood, is a sure way of getting the notes right. I still have the solfa copies of *The Messiah* which I must have had for some performance early on. Alex played the hymns and psalms in solfa in four parts with dreadfully involved key changes for years before he taught himself to read staff notation. This was a singular feat considering that the notes did not correspond to recognised intervals on the keyboard. Had he not been getting it correct, the singers would have been quick to point that out.

My mother detected early signs of Alex's musical talent when, as a very little child, he would push glass tumblers into the sink to hear the sound they made. She told me this when I remarked on the few patterned tumblers that were left of her wedding gifts. My parents could not afford to send him to a music teacher. Having derived great pleasure from playing the piano and organ for most of his life, he at last took lessons in Glasgow, but by this time his eyesight was poor and he had difficulty in reading the music.

Our living room had a blackleaded and steel range, as well as a double bed and later a bed settee. It was never meant for human comfort and certainly gave us very little. Washing which had just been scrougened (partly dried) was draped over a tall wooden clothes horse and this got priority at the fire. We had to leave the sitting-room door open to listen for Granny and I never remember feeling warm in the house. We all suffered from chilblains on our hands and feet as well as mouly heels (chilbains on the heels). The not very effective treatment for these was a green soap-like substance called 'Snowfire' which was certainly no cure, but I think had a temporary soothing effect. My mother cooked and heated water on the range (which burnt coal), until much later when she had a gas cooker installed in the shed. She cooked on this, running in and out with the food. Sometimes in the summertime she set a table in the shed and we ate there 'for a change'. Our house, in the fashion of the time, was lit from a single gas bracket attached to the mantlepiece. A fragile mantle inside the glass globe produced the light which was quite good in limited space. Mantles were made from fine gauze, easily damaged by the touch of a taper or matchstick. This happened all too often and we were kept running for new ones. Most shops sold them, but they cost money so we tried to be careful.

My parents slept in a little back bedroom which had a double bed, wardrobe and dressing table all in close proximity. A tiny window faced Society Street. When my grandparents slept there, an old lady had a habit of tapping on the window to ask 'Whit's the time on the knock (clock), Hugh?' This window was also used to pass on news of deaths in the community, often in the early morning. I remember this happening when I later slept in that bedroom.

My mother bathed us at the fire in a big zinc tub. Then we got dressed in combies (combinations), all-in-one garments (wool or cotton according to the season) which buttoned down the front and were used by both boys and girls. This, over a vest, was a warm foundation for our other clothes which included a liberty bodice, tweed skirt and knitted jumper for girls, and a shirt and jumper for boys. Boys until their teens wore short trousers and knee-high stockings. Going into 'longs', a sign of growing up, was eagerly looked forward to by young adolescents. Our shoes always had tackets (studs) on the soles and heels and we needed them to save wearing down the leather, as in winter we slid all the way to and from school on ungritted roads, and skipped for hours on the dust and metal roads outside the house.

As we grew up, we were handed down clothes from older cousins. I don't remember having a new coat until I was about thirteen. There were always navy-blue trench coats in the house and they were worn by either sex. We wore knitted jumpers and stockings. My mother told me that she would knit a pair of socks for one of us after we were all in bed at night. The women in the fishing community were well used to knitting, as jumpers, drawers (under garments with legs, for the lower body) and socks had to be made for the men's work. Fine ones were knitted for 'dress'.

My mother had a secondhand Singer sewing machine on which she made all her own curtains, and clothes for me. It was a treadle machine, the works of which retracted when not in use. She had bought it when she was eighteen, probably with money from a gutting season. When we were children we used to run the machine furiously for fun without ever using it for sewing. The easily manipulated pedals enabled us to imagine we were doing all kinds of daring things. Despite our use of it as a plaything, it still worked adequately as a sewing machine and was given away many years later as a working utensil. I learned machining on it as a young girl and made a number of garments, one of which I greatly regret making. This was a dressing gown made from my Granny's Paisley wedding shawl. In retrospect this was a terrible misuse of a precious heirloom, but at the time it seemed to be the only material available and I badly needed a dressing gown.

My Auntie Jessie (Granny's sister) used to let her house during the summer months and one of her visitors was a teacher from Falkirk. She had two nieces who visited Nairn with her. We got to know them quite well, especially Miss Mathieson who showed an interest in our education.

Now and again a large parcel would arrive at Auntie

Jessie's; it contained some of her nieces' cast-off clothing. My cousins and I, for whom it was intended, had a lovely time going through it all and selecting a few items that might fit or suit us. This augmented our meagre wardrobe and made us feel different for a little while.

Auntie Jessie was still doing her rounds as a fishwife when we were children and she used to bring home large bunches of flowers for us to take to the Congregational church flower service. We attended this service with our cousins, whose parents were Congregationalists, once a year. I recall wonderful Canterbury bells, lupins and sweet williams which we proudly carried to church.

My mother complained quite often about the lack of space or convenience in the house. It must have been frustrating never being able to clear the living room for comfort. Everything was done in the one room: cooking, eating, washing up, homework, playing, sleeping, and meeting friends and relations. The table was in the middle of the floor, with chairs around it. But most important and overriding all this was the closeness of all the family which was comforting and companionable. Although poor, we were always provided for, loved and cherished. We were secure and happy at home, with Granny at the other end of the house.

Alex and I used to carry Johnny round the house in an upturned stool. I think that he enjoyed it, but he twice fell off and dislocated an elbow. I don't remember him crying or making a fuss, but my mother was very angry the second time, having warned us after the first accident. At the time there was living in the countryside a famed bonesetter at a farm called Balnagown some two or three miles west of Nairn. This man was actually an animal bonesetter, but people preferred being treated by him rather than at the hospital, probably because a well-liked doctor of the town, Dr Eric, had had a broken leg, was treated in hospital and was left with a very pronounced and troublesome limp. For whatever reason, my parents took Johnny to 'Balnagown', as the man was commonly known; they probably walked there. The bone was set in a thoroughly satisfactory manner. It was not Johnny's last visit and he later described how 'Balnagown' massaged his elbow with warm olive oil until he thought it was lubricated enough, then our father held him round the waist while the bonesetter pulled up his forearm until the joint was relocated. This part of the procedure was very painful, but 'Balnagown' was a very gentle and understanding man and his attitude greatly ameliorated the temporary suffering. What a modern osteopath would make of his methods hardly bears thinking about, but many people were grateful for his ministrations. The bonesetting abilities seem to have been handed on, because in addition to 'Old Balnagown' there was a 'Young Balnagown'.

When my father was home from the fishing in winter, a favourite winter occupation of his, and a necessary one too, was the making of thrum rugs for the fireside. Thrums were the ends of woven tweed which he obtained cheaply from the Holm mills in Inverness. These ends were where the weaving had stopped and only the warp, with an inch or two of the tweed, remained. The long strips of the warp were bunched together to about one inch in diameter and cut into lengths of about two inches. We used to 'help' with this job. The resulting tufts were then stitched onto sailcloth which formed the base of the rug. The closer each bunch of wool was, the more luxurious the result. My father often used a sailmaker's palm, a leather protection for the hand which helped to punch the needle through the thick, tough sail-cloth. Although the random colours from the thrums were perfectly attractive, we children used to draw patterns for him to follow. They were not usually very imaginative, being mainly geometric shapes, but he tried to follow them to please us. These mats, as we called them, were very useful and cosy under our feet and we usually had at least one new one each winter.

Another household activity of my father's, which fasci-nated us, was the making of churls. These were very effective home-made firelighters probably originating from the days when people did not have newspapers to ignite the kindling. The churls were made from small sticks about six inches in length and an inch and a half in diameter. The technique was to shave the wood with a sharp penknife (which fishermen always had), from about a third of the way up the stick to about two thirds of the way and then bend it over without severing the sliver. The result looked like ruffled feathers and was very attractive looking. But, more important, the churls were invaluable for establishing a good blaze in the fireplace very quickly.

In the mid-1930s, when the herring fishing was being abandoned by my father and many others in Nairn, my mother's brother, Uncle Isaac, a trawlerman in Aberdeen, gave my father a berth in his boat the *Newhaven*. Relatives moved in to stay with Granny in our absence and we all went to Aberdeen for about two months one winter. Alex, now of secondary school age, attended the Central School on the recommendation of Dr Milne, rector of Nairn Academy. Johnny and I were enrolled at Walker Road School in Torry. It was a huge primary school compared with Millbank which

itself had over five hundred pupils. I remember being surprised to see the children being given small bottles of milk at
the morning interval. I did not partake of this, so I presume
that it was a welfare issue. I had an easy time at this school as
I had already covered the stages in Maths and English that the
class was at. My teacher was Miss E. D. Cook who was a very
fine person. My school record had this note from her: 'This
child has made 78 attendances out of a possible 78. She is a
very good little pupil and has done exceedingly well in all
subjects'.

Our stay in a city opened our eyes and broadened our
horizons a little bit. We went to the cinema and the market,
where we watched with amusement as people queued up to
buy lunch at the restaurant; I well recall the counter staff
calling out to the kitchen, 'Pie and chips one', 'Sausage and
chips two', etc. We ate their pies and chips too and had a
good look round the whole place. We used to buy boiled
haddock roe in firm little pieces which we could eat on our
walkabout. We went once to the local cinema where we
watched the film *Bonnie Scotland*, twice around. It starred
Shirley Temple with her 'father's' regiment on the North
West Frontier in India and we were thrilled by it. My mother
was not so thrilled and was very worried by our long
absence.

We visited Duthie Park quite a lot when we went to see our
relatives living in that area. Our great interest at the time was
to fish in a pond there for tadpoles and bannies (sticklebacks),
and we had bamboo canes with little nets at the end for
catching them. Many people sailed large toy yachts on the
pond.

Our shopping was done at the nearby Co-op. We were
greatly captivated by the idea of the 'dividend'. How much

we benefited from this I don't know, but we were always glad to get 'something for nothing'. After making our purchases, we gave the assistant our numbers and got a copy of the ticket on which it had been recorded. All this was then transferred to a passbook which showed our dividend at the end of a period. We received cash back, based on what had been spent.

We attended a very good-going United Free church in Torry. This was home from home and we loved the hearty singing. It was attended mainly by the fishing community of trawlermen who were prospering at that time. We were fascinated by the number of ladies who dressed in elegant fur coats.

Our accommodation for the two months was a single room in a tenement, sublet from a tenant, in Menzies Road. There was a bed in a recess which probably slept three of us; I cannot remember what other bed we had. Into this frugal accommodation we welcomed quite often a Nairn cousin who was doing a nursing course at the City Hospital. She was glad to spend some of her days off with us, even managing to sleep over on occasions. Some relatives of my father had found this room for us; they had the small flat at the top of the tenement. They were childless, and we spent a lot of time with them as they were very kind and often took us out.

This was a happy time in our lives. My mother felt liberated and was pleased to have a small but regular income for a short period. We got a little pocket money for the first time and I bought a schoolgirls' magazine called *Girls Crystal,* Johnny bought the *Dandy* or *Beano,* and Alex had the *Hotspur,* the *Skipper* or the *Adventure.*

10

SCHOOL

The infant school at which we all began our education was just a short distance away from our house. It had provided for all the Fishertown children, and some from surrounding areas, since 1872 when the Education Act made school attendance compulsory to the age of 13. In my time it catered only for children aged 5-7 years, but my mother and her contemporaries remained there until early adolescence, getting a secondary education which included Domestic Economy and French. I think they were well taught, being competent in the 3Rs and making good use of them all their lives.

My mother always spoke warmly of the 'Maister', Mr Richard Jamieson, one-time head of the Links School. His wife also taught there and after them two of their family, Grace and Muriel Jamieson. Muriel taught my brothers and me for the first two years. I remember this as an idyllic time of my life. I found out later from Miss Jamieson when I myself was training to be a teacher in England, and was telling her about my Froebel course, that she, in her early years as a teacher, attended Froebel courses in her summer holidays. Froebel, a German educationalist, believed that 'play is the child's work'. He advocated 'learning by doing', and Miss Jamieson adopted and to some extent at least practised methods in keeping with this philosophy. This helps explain the happy days of our early education. In our classroom there were playthings of various kinds, including a large and beautiful dolls' house furnished in the 1920s fashion and peopled by little jointed figures in clothes of the same period. Members of staff at the school used to bring back from holidays small pieces of furniture and utensils to help furnish the dolls' house. It is now on loan to the Nairn Museum from the Education Authority. There were also coloured charts on the walls of our classroom. I remember especially one about numbers. There was a big sand tray and water play in another room. We used slates and slate pencils for writing and sums. The slates were taken home at weekends to have the wooden frames washed and we had a sponge for cleaning them in an Oxo tin with a little water and Izal in it. I learned to read early. The *Daily Express* was one of the most popular newspapers of the day and I loved the stories and pictures of Rupert Bear written by Mary Turtle. I took a cutting every day to school and stood beside Miss Jamieson reading it to the class. At the end of my infant schooling I was awarded first prize equal with Jean Leith. We had to go to Millbank School on prize day (my prize was a big book of nursery rhymes) and I remember being quite disorientated at the other end of town. I had never been there, yet after the summer holidays I had to take a long walk to and from school twice a day. We had an hour for dinner and always went home for it, contriving never to be back late for class. At the Links School on days when lessons were successfully completed, one of the children was sent to the nearest shop for a tin of biscuits and then we all enjoyed a tea party in the schoolroom.

One day, when I was six and suffering from a severe chest cold, I was reluctantly taken from the box bed by my mother and dressed to go to school. An Inspector was coming and

Salt Herring on Saturday

Miss Jamieson had asked that I be present if at all possible. When I was seven and in Millbank School, I was moved with a few other children to the class above, so skipping a year's education. We were placed with Miss Ross, a teacher who already had a class of forty, and she used to gather us in a little group round her table, and give us extra tuition to bring us up to the level of her other children. The experiment, if such it was, seemed to have worked, because I was not aware of any problems which might have resulted. Strangely enough, I ended up where I started, with my own age group, because of an illness at the start of term when I should have begun secondary education. I had been to my Granny Bochel's and been made, by an aunt, to look at her dead body in the coffin. On the way home, which was only about two hundred yards, I stopped walking as my legs became very sore. I must have managed to get home, and when the doctor came he diagnosed acute bronchitis. This kept me at home for about eight weeks and it was obviously pointless to enter Nairn Academy where my class would have been well into Latin, French and Mathematics, and I couldn't hope to catch up with them. So I returned to Millbank to repeat a year.

One summer day, when I was about nine years old, we all went, with cousins and others, on my father's steam drifter *Glenerne* to sail along the Moray Firth and through the Caledonian Canal, disembarking at Dochgarroch. From Dochgarroch we took a bus back to Inverness. It was a beautiful day with warm sunshine. The sea was calm and we could see large groups of jellyfish floating near the surface. The drifter and its crew were on their way to the summer fishing on the West Coast and my brother Alex was to take his first trip with them. He was about eleven years old. It was customary for boys of this age to experience life at sea, but the main purpose was to give them a holiday. Alex said very little about it when he came home and never considered, nor would he have been encouraged to consider, fishing as a career. When we got to Inverness that day, there were crowds on the pavements watching a regiment of Scots Greys marching along with their band playing. It was an impressive sight which I wrote about in a composition next day at school. For once I had plenty to write about and was very pleased with myself until the teacher read out an extract and stopped when she came to my description of the march. I had described the soldiers as the 'Scots Graze'. I think she got a laugh out of it, but in my innocence I had just put the rules of phonetics to work. I was hurt and have never forgotten the incident.

Once in Millbank School when I was in a senior class, we had our first radio lesson. We were all seated with pamphlets on our desks to listen to a nature lesson on robins. Johnny's memory is of having the radio in the hall, full of children who, despite the sanctions, couldn't keep completely quiet, so the children couldn't understand a thing. It was a complete failure. The reception was hopeless, the acoustics abysmal and even to our youthful ears the speech was indecipherable. The teacher gave up and we never saw the radio again. We were sorry because we thought that this would be a fine change and perhaps a bit of a skive.

11

SECONDARY EDUCATION

Nairn Academy was a secondary comprehensive school, and when I started there after the summer holidays in 1936, I immediately loved all the lessons. It was a complete change spending forty minutes in each class and getting a different teacher each time. There were country children who cycled in to school and often got wet and very cold in wintertime; on arrival at school they were sent to get dried off and warmed up in the boiler room.

On my first day in the Latin class, the teacher, Miss Linton, walked up and down the classroom repeating the word 'ambulo'. She wanted us to say 'I am walking', but none of us had the wit to see what she was getting at. However, we soon picked up the basis of Latin grammar and enjoyed the declensions of verbs and the conjugation of nouns. I entered into it with great enthusiasm and used to hurry home after school to open my books and get on with some exercises. I think that it was the orderliness of the language that appealed to me, and the rules which, once learned, were easy to keep. One could never forget that the verb came at the end of the sentence and that when nouns were plural, adjectives were plural too. The top class in Latin was taught by the Rector, Dr Milne. Although he was austere in manner, I always found him relaxed and amusing in the Latin class. We used several textbooks that he had written and I believe they were used throughout the country. A relative who was educated in Lancashire remembered his *Easy Latin Readings* and *Latin Prose Composition*.

As my parents discouraged us from doing homework on Sundays, I used to go upstairs and work on the (very small) landing to do revision. Later, though, after Granny died, we slipped into the habit of taking out our schoolbooks on Sundays and my parents made no objections. We never gave the teachers or school authorities any trouble, but when Dr Milne insulted Alex in front of his class one day because my mother stopped him from taking swimming lessons, she put on her coat and hat and faced the Rector about it. She had spent much time and care nursing Alex through asthmatic attacks and felt that she had the right to decide whether or not he should attend the baths. He was a quiet, self-effacing boy who would have been deeply hurt over the incident. My mother insisted that Dr Milne apologise to him in the class-room with the teacher and all the pupils present.

I was astonished to find in our house during my Academy days a Latin grammar which had been used by my uncle Hugh Bochel. He was born in 1880 and must have been one of the boys sent from the Links School to Nairn Academy in the late nineteenth century. As the years went on, I found that some of the subjects palled. I enjoyed and did well at French, English and History, but Maths, apart from the arithmetic, began to elude me in the upper classes and I settled to concentrate on arithmetic in my final two years.

I don't think that any of us as a family got concerned about exams. We were never harassed at home about our school-work and sat exams as a matter of course, even looking forward to them as a change to the daily routine. Alex took a very good group of Highers which after the war gained him

immediate entry to Aberdeen University to study Civil En-
gineering. I had an inferior group which would have been
sufficient to get me into the University, if family finances had
permitted; later in life I was able to progress into higher
education. Johnny was less keen on school than Alex and me
and left at 15 to serve an apprenticeship as an electrician. His
great interest in politics later led him to the University of
Manchester to study Politics and Modern History and he
ended his career as a Senior Lecturer in Political Science at the
University of Dundee. Psychology was not a strong point in
the running of Nairn Academy. When we were waiting for
the result of our Highers, the Rector announced them in front
of the whole school at assembly time one morning. For the
first time I became anxious and felt as if I might faint, until
my name was called out as being successful.

There was an attempt at real living when dances were held
for different stages of the school and the very good concerts
which were organised at the end of each session. There was
excellent choral singing by all the classes, plays and dancing.
Most parents came to the concerts and the money raised
helped to boost school funds.

After sitting the Highers, girls were given cookery lessons
by Miss Cumming. This was a brilliant idea which not only
let us wind down from our academic work, but was both
enjoyable and useful. In the short time we had, I learned some
things about pastry- and cake-making that I have never
forgotten.

I was picked for the first eleven at hockey, but managed to
play for the team only once, as I was given a Saturday job in
my uncle's grocery shop and this took priority. I was dis-
appointed but had to give way, as the five shillings I earned
was very welcome to the family.

Secondary Education

Most of my friends were going on to Aberdeen Training College to become teachers, but without a bursary this was impossible for me. I later trained in England and later still became interested in children with special needs and qualified to teach them. I left school at sixteen in 1941 to work in the local Food Office. A system of rationing food supplies to the public, to shops and hotels, had been set up and was well established by this time. We issued ration books and identity cards. People moving from place to place were required to register their new address and get their documents amended. New babies when registered were given ration books and identity cards. I remember a lady coming to the office to notify a change of address. Her previous place of residence had been 'The Battlefield', and now she was moving to 'The Retreat', in Nairn.

I always admired my parents' skills in the three R's, but now realise that they were very well taught and thoroughly immersed in the crucial subjects. It explains why they had no difficulty in putting pen to paper later to write many letters to relatives abroad and to us when we left home.

In the Links School in their time of the 1890s and early 1900s, as well as the basic subjects, older boys had lessons in Latin, algebra and Euclid, agriculture, advanced arithmetic and composition, 'but the want of books is a great drawback' (from the Links School Logbook). In 1898 it was reported that 'composition is very defective owing to the want of books in the homes of the children and the character of dialect used by parents in conversation'. Small amounts of money were given by the Board for the purchase of books in school and for the most part the best-performing pupils, but access to books was wholly inadequate.

Swimming lessons were begun in 1902 and payment made

by the Board for twelve boys. Nairn swimming pool was then situated where Marine Cottages now stand. It was privately owned and run. It was thought, quite rightly, that the swimming helped discipline. Perhaps so, if you were not pushed backward into the water. Mr Jamieson, 100 years ago, was taking senior boys for nature walks along the seashore. They noted 'The sandstone in the quarries, the boulders left by the ice in the glacier period, the wild flowers and the great standing stones that marked a clan battle'. The boys enjoyed the walk and expressed a wish to see the local museum run by the Nairn Literary Institute. On another walk 'They gathered hop, bird's foot trefoil and other plants of recent nature knowledge lessons. Different varieties of seeds were noted'.

Domestic economy, French, religious education, music and drill were on the curriculum when my mother was a pupil at the beginning of the twentieth century, but there was some difficulty in getting girls to bring materials for sewing, and knitting had to be substituted. The domestic science subjects must have been thoroughly taught because my mother, all her friends and female relatives were well versed in them and used them to the benefit of their families when they got married.

In September 1895 Miss Macdonald was appointed to conduct cookery classes for girls of upper standards at a salary of £10 per session, 'the teacher paying the incidental expenses'.

A full-time teacher's pay at the beginning of the twentieth century was £50 annually for those newly started. She would be in charge of 150 pupils and have three or four pupil teachers to help. An inspector in 1905 reported: 'It was wrong to expect one certificated teacher, however many PTs she may have for part of the day, to educate about 150 children'. In 1895 a male teacher was appointed at a salary of £80. The school roll then was 133 boys, 104 girls, and Infants 41 boys, 55 girls.

Grants depended on attendance:

	£	s	d
On average attendance	103	0	0
On girls average attendance		3 16	0
On average attendance			
(Scholars under standard III)	10	5	0
Grants for pupil teachers	4	16	8

In my parents' time, attendance at school was very desultory. Various seasonal pursuits, and helping to prepare the boats for sea, meant unauthorised but understandable absences, especially of boys, who were pretty rough and needed a lot of discipline. In winter time the quarry pond (where the putting green now is), when frozen over, attracted them before and after school and also at dinner times. Mr Jamieson lived in Carron Villa near Viewfield, and used to pass by when most of his pupils were revelling on the ice, but instead of harrying them, he took no notice and went ahead to ring the bell. In June of each year, Nairn drifters went to Fraserburgh for the summer herring fishing. Older boys, along with luggage and household equipment, sailed in the boats with their fathers, while younger children accompanied their mothers on the train. Money laid by in the Penny Savings Bank, begun in the school in 1899, was used mainly to pay for train fares to Fraserburgh and sometimes for boots.

Home was transferred as completely as possible for about two months to this busy east-coast fishing port. Granny and her family stayed in the same rented accommodation in

Charlotte Street every year and the children rejoined old friends for games and church activities.

This annual migration was the main cause of poor attendance at the Links School. On 6 July 1894, the Links School logbook recorded that 'Attendance was much reduced because the fisher children left much earlier this year'. Poor attendance was a great problem and seriously frustrated attempts to fulfil curriculum demands. The older boys played truant a lot even though checks were regularly made by attendance officers. Other reasons given for absences from school included 'carelessness of parents, idling at home, market days, caddying at golf, bramble picking, weddings, severe weather and selling firewood'. On 25 January 1895 it was reported that 'Weather is extremely cold and attendance in lower classes is very much reduced. Many of the children are ill fed and poorly clad, and some sicken with the cold in school'. That same year the use of heating stoves was sanctioned and, later, grates were installed in one room holding 150 or so children.

Lack of hygiene in homes at this time resulted in severe epidemics of childhood illnesses such as measles, mumps, chicken pox, scarlet fever and diphtheria. Few children were left attending school at these times, and in order to prevent further spread of disease, the Links School was closed a number of times for periods of three or four weeks, as on one occasion 'the epidemic of measles and mumps had increased to an alarming extent and several deaths had taken place'. On 5 February 1897 'A severe storm of snow and frost prevented many of the children from being at school'.

Households where there was no running water had to make do with using a standpipe. There was one just across the road from our house. Some houses had outside water closets in small sheds, but there were some without even this facility.

Defaulters from school were regularly visited by attendance officers and parents were sometimes prosecuted, after many warnings, paying a fine of ten shillings, a large sum in those days and equal to half a week's pay for a teacher. However, even this did not make much impression. So many absences meant that work had to be repeated often, and this held back progress and demoralised the staff. Inspections by the School Board were carried out frequently and suggestions made for improved performance by the pupils. Sometimes reports were good, but always urged better attendance.

Pupil teachers were engaged both for assisting in the huge classes and also as part of their training to become certificated. Grants were made to the school for this. In 1895 the grant was £4:16:8. These young women, and sometimes men, also studied during this time of training and sat weekly exams to qualify for entry to one of the Church of Scotland colleges in Aberdeen or Edinburgh. They were regularly assessed in arithmetic, grammar, composition, history, geography, algebra, science, hygiene and chemistry. In 1900 pupil teachers were recommended to read *The Spectator*, *Macaulay's Essays* and Southey's *Life of Nelson*. So they had a fairly wide experience of study and a great deal of teaching practice before going to college.

One of the pupil teachers at the time was David Fraser who later went to Aberdeen Church of Scotland Training College and was still teaching in Millbank School when I was a pupil there in the 1930s. He was headmaster and known as Major Fraser, having reached that rank in the army during World War I. He took on the formidable task of teaching grammar to all upper primary classes. 'Major's grammar' was a renowned and feared subject. It was fine if you could do it, but was a terror to those who were either not interested or unable to absorb it. 'Can I get a copy of Major's grammar?' was a frequent request at our door in those days. I was fascinated by the different tenses and wondered what relevance they had to our lives. Later in the Latin and French classes in the Academy I found out just how useful and necessary they were. When the upper Primary school classes were having swimming lessons in the Nairn pool, 'Major' paid an annual visit to see how they were progressing. It was his custom to line up the whole class with their backs to the water, and push them in one by one. This was a terrifying experience for those who could not yet swim and we very much feared and resented it. It had the effect, at least on me, of putting me off swimming and I never learned to enjoy the water or become a confident swimmer. My brother Johnny never did learn to swim and he attributes his fear and dislike of being in the water to this experience.

The Links School was a precursor of what was later known as a Junior Secondary School where primary and secondary pupils were taught in the same building and under the same Head Teacher. The Junior Secondary survived in Scotland until the 1970s when comprehensive secondary education took over. Nairn Academy was always comprehensive, taking pupils of all abilities, but sorting them into groups for Latin, Commercial and Technical.

12

THE GUTTING

My mother was one of the fisher girls who grew up to become gutters and packers of herring and carried on this trade until just before marriage. They went to Lerwick, Fraserburgh, Yarmouth and Lowestoft, following the fishing fleet. The seasons were short but intensive and badly paid, but the girls loved living and working together, and being away from home for eight or nine weeks, with the company of the menfolk at weekends. Courtships flourished and Seaman's Hall records show that marriages took place mainly in the intervals between fishings.

The years before World War I were the busiest for the herring industry. Large fleets of boats were needed to satisfy the enormous demand for salt herring from Germany and Russia, but this trade came to an end after World War I because of restrictions put on exports by the Government.

My mother hated sailing to Lerwick from Aberdeen. It was often a rough crossing lasting for about 16 hours and the passengers who slept on deck were often very sick. Even experienced deep-sea sailors were affected by the rolling and tossing of their ships in the Pentland Firth and the dreaded Sumburgh Roost.

The girls were engaged for a gutting season by teams of men from the curing side who visited all the places where gutters and packers were recruited. The girls arranged their own crews of three: two gutters and a packer. These crews remained intact for many fishings. Only illness or marriage broke up a partnership. Close friendships were made which endured throughout their lives.

Accommodation for the girls was in wooden huts close to the curers' yards. One room, with two bunk beds each sleeping three people, and primitive cooking facilities (a fireplace) were provided. A table was the only other piece of furniture in the room and the girls sat on the kists in which they had brought all their domestic utensils from home.

There were no prescribed hours for gutting. When herring were landed, the girls worked until they were all cleared, however long this took. They were prepared for anything, the big rush caused by a glut of fish, when the work might last from 6am until 10pm, or the lull between catches, when there was nothing to do but knit. The girls were not happy idle, as a busy life of gutting for them meant their menfolk were doing well too.

There were never any grumbles about housework. Each girl had her 'day on', when she was responsible for rising first in the morning, lighting the fire, fetching water and cooking all the meals for the two crews who shared the hut.

A cooper banging on the door with a stick awakened them at 5am, calling out 'Get up and tie your fingers!'. Before going to the yard at 6am to start 'filling up', they had tea with bread and butter, and tied up their fingers. Breakfast at 8.30am was porridge, bread and jam. Work on freshly landed herring started at 9.30am.

The fishergirls' hands, which had to work at high speed, were protected from salt and the sharp gutting knife by cloth bindings tied round each finger. Each girl did her own, using her teeth to help secure the knots. Forefingers were the most vulnerable and needed special attention. Despite these precautions, injuries to the hands were common and nurses were in attendance at missions, to apply dressings.

The gutters worked from farlins (rectangular wooden troughs), with up to six tubs arranged around the backs of each pair, to hold the different selections of fish. They were sorted into smas (smalls), matties (young and maiden herring), matt fulls (sexually mature herring), fulls and large fulls. A girl could gut and grade 60 to 70 herrings a minute, 20,000 a day. Each barrel held between 700 and 1000 herrings, depending on size, and a packer aimed to fill three barrels an hour. Gills and gut of the herring were removed at one stroke of the short, sharp gutting knife held between the thumb and forefinger. Melg (milt) and ran (roe) were left in. Herring were partially salted in the farlins, and salted again as they were packed into barrels. Packers followed an accepted arrangement of packing at which they worked swiftly and skilfully. Bottom and top layers were designed to show the herring at their best. Buyers sometimes tore herring from the tail upwards to taste them and the Russians opened barrels and tipped them up to see if the herring were all the same size.

The girls worked as hard as they possibly could in harsh conditions to keep their earnings up, but they considered £17 to £20 a good income for a season.

In spite of their strong attachment to Gremista in Lerwick where they experienced freedom, friendship and fun, the girls were always glad to be going home, away from the harshness and exhaustion of their working days. Many knew that this was certainly their last visit. Marriage and home-making brought their travels to an end.

The kind of partnership created by marriage within the Nairn fishing community was, at the time, essential for the successful operation of the local herring fishing fleet, which was heavily dependent upon the skills and co-operation of the fishermen's wives who were required to mend nets, knit sea guernseys, drawers and socks, and undertake heavy washings. They also packed the chests for the fishing trips and prepared bedding for the ships' bunks. This was a mattress filled with chaff commonly known as a caff-seck, and layers of blankets and quilts sewn together as a covering. The wives of fishermen had to accept their husbands' long absences from home, the dangers of the job, and the prospect of bringing up their families alone for months on end with no guarantee of a regular or adequate income.

The Seaman's Hall was a fine place to have weddings, and after it was built in 1887, all the men and women in the community were married there, the ceremony being held in the Library and the reception in the main hall, where up to or over a hundred guests would be entertained. Ashers, the long established bakers, provided the meal which was normally steak pie with mashed turnips and potatoes and a choice of apple tarts, jellies and trifles for dessert. This was followed by tea and wedding cake.

The hall was then cleared for dancing and the committee room set up with refreshments for the intervals between dancing. There would be several sittings for the tea served by the bride's friends. The bridesmaid then distributed conversation lozenges, each man choosing one for his partner, trying to take one with a pleasing motto.

My mother bemoaned the disappearance of her wedding cake, which had been consumed by a group of naughty boys when everybody else's backs were turned.

There was no honeymoon. The bride and groom were accompanied to their new home by the young men of the community who spent the rest of the night playing tricks on them and emptying the house of food.

For her home a bride was expected to provide blankets, quilts, sheets, towels, pillows and bolsters, pots and pans, a brush and shovel, ornaments, pictures and a mattress for the bed. Her family paid for the wedding cake and the groom for the reception. Mr John Ian's bill for his wedding dinner for 172 guests in March 1930 was a little over £30. After the 'kirkin'' (churching) at the bride's church the next Sunday, the young couple began regular attendance at the husband's church, often the same one.

For the first time, in 1909, a printed invitation was sent out for the wedding of Mr and Mrs David Mackintosh of 24 Society Street. This specified 'young couples in strings as usual', meaning that everyone was expected to march in file to the Seaman's Hall. Before this, 'bidding' to weddings by word of mouth round the doors was the custom, and this was usually preceded by a ceremony called 'the contrack' (contract), when close members of the two families met in the bride's home and witnessed the formal betrothal of the couple, and application was made to the session clerk of the parish for proclamation of the banns.

My parents were married on 23 December 1921. Eight marriages took place in the Seaman's Hall that year. There would perhaps be two weeks before the fishermen set sail for the winter fishing on the west coast of Scotland, and many preparations had to be made for this, including packing his kist.

A fisherman's kist had a shottle (small hinged box) attached inside. Small items were put in it so as to be easily found. These would be tobacco, tie, handkerchiefs, pen, paper and ink, stamps, purse, spare bootlace, cut-throat razor

in its case, shirt collars and bible. The kist itself contained
Sunday clothes for the church and visiting. Items of Sunday
dress were a three-piece navy blue suit, a fine knitted navy
blue jersey and a Sunday bonnet made of soft tweed with a
snood. There were also spare socks, a towel and a pair of fine
black boots. Working clothes were packed into a bag, the
underclothes being rolled up together in sets. Bags of washing
were sent home at intervals to be washed, except from
Yarmouth, when everything was brought home at the end
of the season, which meant a very large and heavy wash for
the women.

13

PLAY

Houses were built close together in the Fishertown as land was in short supply. Only a narrow strip of ground about two feet wide separated those set with their gables to the road. We called these little tunnels 'skuttlies' and they featured prominently in our play as hiding places. In the dark winter nights some children would not venture in too far so the hider was safe. Today skuttlies are closed in at one or both ends by corrugated iron or some other material which is removed when repairs are required at the backs of houses.

Groups of children played around the 'lampie', in our childhood days a street lamp lit by gas. 'Our' one, Willie Barron's lampie, was conveniently placed at the crossing of Society Street and Union Street and it was there that we gathered to decide on the different games to play. 'Kick the tinnie', 'Alease' (release) and of course hide and seek were the ones I remember. 'Kick the tinnie' was a form of hide and seek, but instead of counting 5, 10, 15, 20 . . . the can was kicked as far away as possible and the chaser had to fetch the can back before pursuing the rest of the group who were hiding. If the hider got back to the lamp-post without being touched by the pursuer, he or she was safe.

'Alease' was also a version of hide and seek; the variation this time was that if a hider was caught, he or she had to stand by the lamp post; if another hider could get to him or her

71

undetected, the captured one could be 'released'. One interesting feature of our play was the sense of fairness that usually prevailed; of course there were no disputes, but I can recall few occasions when the equivalent of 'taking my ball home' was resorted to. There was also an understanding about territories used in hide and seek; it did not seem to be necessary to define legitimate distances from the lampie for hiding.

If the weather was dry, we gathered every night to play and make plans for next day or compare notes on homework. As the Fishertown was densely populated, there were always other children of our own age. Boys and girls played together. We must have been an awful nuisance to the neighbours at times, but they were remarkably tolerant. Sometimes a 'bobby' would appear as if by magic, hastily writing something in his notebook, but nothing ever came of it. For we were law-abiding, strictly adjured to respect other people's property. Besides, the older people round about got their own back by sending us on errands during the daytime. I used to think this unfair, but we often got a penny for our trouble and that was very welcome.

Although we had a wonderful stretch of links quite near, we played mainly in the street. Somehow this was more companionable and there was very little traffic at the time. For skipping in groups we had strong heavy ropes from our fathers' fishing gear, which, when 'cawed', lashed the road with a heavy smack. Two children cawed while others, sometimes quite a crowd, queued to skip in and out of the rope. We changed over often so that everyone got a fair turn of cawing and skipping. Democracy ruled. Those who demurred were sent on their way, and were more co-operative next time they came to play. Games had their seasons. When the brighter days arrived about March, out came the ropes, the balls, the 'boolies' (marbles) and the girds (hoops from barrels or bicycle wheels bowled along with a stick). For some reason boolies was mainly a boys' game, but girls did sometimes play. The name 'boolies' probably had a connection with the French game of boules whose balls were at one time made of marble or another stone that could be fashioned into a perfect round shape. There were several games played with boolies; the first was often played on the way home from school in the strannies or drainage gullies on the High Street. One player threw a boolie forward and the others took it in turn to try and hit it; the first to hit it won the bool.

Another game was played on the many dirt closes or lanes in the Fishertown; this time a shallow hole was carved out of the soil using the heel of the shoe or boot (shoes were much abused in our play); the hole was shallow so that the game was not too easy. The object of this game was to get the boolie in the hole from a predetermined distance. David Ellen described to me two games played with boolies. A circle was drawn on the ground and a bool placed in the middle of it. Each boy in turn, after taking a bearing on the target, dropped his bool. If he succeeded, he won the bool. Isaac Main of King Street excelled at this and thereafter the game was known as Eye-aac. Another game was trying to hit or get closest to a bool some distance away. The boy who won picked up all the other bools.

In our young days the glass marbles that we know today were less common and more expensive than they now are; they were considered quite valuable; more common were stonies (larger than glassies) which were, I think, made out of some ceramic substance, and were probably originally marble. The difference between the glassies and the stonies was like the difference between copper and silver although today the stonies would be infinitely more valuable. There were two standard sizes of both kinds, small and large; the latter were greatly prized and known for some reason as 'Donals'; the smallest were called 'Peachies'. Mothers made little cloth bags with drawstrings for holding the boolies or durbs as they were sometimes called.

Peashooters were the rage at certain times of the year, autumn, I think, when the bents had ripened and the hollow stalks were yellow and hard. We selected the thickest stalks through which dried peas could be blown, but it was more common to use barley. Both varieties of ammunition were readily available because they were staple ingredients of Scotch broth.

The boys made primitive go-carts out of fish boxes and pram wheels, and they could be steered by allowing the two front wheels on a loose attachment to be guided this way or that with a rope held by a driver. Much fun was had with these where the road sloped downhill. After a push, free-wheeling would take us a long way. Scooters were popular too, for the same reason. Spring and summer days and evenings saw children swarming all over, absorbed in play. In our close there were two broken-down houses which were a great attraction. We climbed in and out of the open windows and explored a small garden opposite where we found lots of yellow and black caterpillars on old gooseberry bushes.

In the cricket season, balls, bats and wickets, some of them home-made, were searched out from among the jumble in the sheds. Boys and girls played cricket together. Everybody wanted to bat. How we missed breaking every window in sight, I don't know, but there were very few such misfortunes.

Play

Nairn County Cricket Club had a big and enthusiastic following, so it was natural that children should copy.

The gable ends of houses had only one window or none at all and were just made for all ball play. We threw a ball against the wall while turning full circle or doing something with our hands, like twirling them round each other or clapping and counting before catching the ball. Also bouncing a ball on the ground and swinging a leg over it while singing:

one, two, three a lary
four, five, six a lary
seven, eight, nine a lary,
ten a lary postman

Girls played skeetchies or beddies (hopscotch). A small flat stone or piece of slate was thrown by each player in turn. If the stone landed in the appropriate clear place, the player hopped into each square, picking up the stone on the way. If she put a foot on the line she was 'out'. If she came back clear she got another turn. All games sessions seemed to have been played in good spirit. If there were disputes, they didn't last long. In true Olympic spirit it was the playing that counted rather than the result. There was a good swing, maypole and chute playground at the Links where we spent a lot of time, and in the winter, when there was snow, we sledged down Bunker's Brae until we were exhausted. In March we flew our home-made kites with long tails made from screwed-up newspapers and string or a piece of herring net. A bank on top of the beach was ideal for catching the wind and the kites flew high and free. We made our kites from split bamboo canes, brown paper and string. Paste was made from flour and water boiled together, and it certainly kept the brown paper securely in place and cost nothing.

We sometimes walked about on stilts made from wood or syrup tins. Holes were pierced in the tins, and string looped through, which we held taut, with a foot on each tin. These were not as daring as high wooden stilts, but they were quite enjoyable for the younger and more timid among us.

During the summer season we paid a few visits to an old quarry poll on Lochloy Road where we fished for bannies. These were sticklebacks, which we kept in jars. I don't think they lived long because we didn't know what to feed them on. We used long sticks with little nets on the end to catch them and we loved this fishing game. We also got tadpoles at the same pool. This activity must have been potentially dangerous, but we were never kept from doing it and I don't remember any accidents.

There were also safer stretches of water in the fields on Lochloy Road, which, being shallow, froze over very quickly in winter. We somehow contrived to find skates which were attached to the soles of shoes or boots by clamps. They often came off when we went fast and we went sprawling on the ice. But it was great fun and we all learned to skate a little.

In the summer holidays from school, teenage boys earned some money caddying for golfers at the Nairn Golf Club. They left home quite early, walking along the seafront, and hung around the professional's shop 'to be engaged'. Some days there was very little doing and the starter often demonstrated quite a bit of favouritism. The pay was usually very poor, sometimes as little as ninepence for a full round, sometimes two shillings, and two and sixpence was considered a satisfactory payment. It was not uncommon for the boys to hide until a very poor payer went on his way, but if this was detected by the starter, the guilty ones found themselves ignored by him for some time after. Occasionally a boy would be 'engaged' for the afternoon or next day and sometimes for a whole week, which was a welcome bonus.

Golf bags were less heavy than they are now, but it was a long walk around the course. Most of the players who used young caddies were not very good players and a major part of the job was searching for lost balls. It was easy to get into trouble for not following the flight of the ball and being able to walk straight to it. Before the days of stainless steel heads, some players demanded that the irons be polished with emery cloth after a game. As well as earning a little money for the family, the boys learned the etiquette of the game and not a little about the technique, so that when eventually they became members of the Dunbar Golf Club, as many of them did, they required no formal training to be competent players. The boys' sisters took snack lunches along to their brothers so that they could stay on all day at the golf course in the hope of 'getting out'.

14

LOCAL CHARACTERS

There were a number of characters in the Fishertown, around whom myths and scare stories grew up. One such, for whom we were always on the lookout, was 'the miser'. This poor man was not a miser at all, but as he lived in what we considered to be miser-like conditions in a but and ben in Society Street, the nickname seemed appropriate. He was a tall man who, when he did appear, was always, summer or winter, dressed in a long khaki-coloured raincoat. He shuffled along, avoiding everyone's gaze, to his house with its cobweb-covered windows through which the light of a candle glimmered in the dark. It was said that he had been involved in the notorious game of cards in which the future Edward VII, as Prince of Wales, was suspected of cheating. Our 'miser' was believed to have taken the blame on behalf of the future king and was exiled from high society, and came to live in obscurity in the Fishertown. It was rumoured that he was an estranged member of a landed family in the county. We never heard him speak so I don't know whether he had a posh accent. I don't think any of the children tormented the 'miser', nor were we afraid of him; the same could not be said of some of the other characters in the Fishertown, although I'm sure that they were all harmless.

An old lady called Nana Callie was often seen leaving her thatched cottage in Society Street in the early mornings, heading for the beach with her lug spade and pail, supposedly to gather lug worms for baiting the lines. When she died, a locked kist was found under her bed by relatives who were unable to move it because of its great weight. When it was finally dragged out and forced open, they were amazed to find it filled with sovereigns which they quickly distributed amongst themselves, each getting a bowl filled to the top and levelled off by their hands. The origin of the treasure was never discovered but it was thought to have come from a wrecked ship and brought to her house in relays in the lug pail by Nana under cover of darkness. Another story is that an officer on a white horse came into Nairn to collect the pay for the garrison at Fort George and was thrown from his horse and killed on the return journey along the beach. The saddlebags with the money disappeared on the flood tide, and from time to time the coins, mainly sovereigns, were picked up by the fisherfolk. In any case, the story of Nana's treasure spread like wildfire, and people were seen running from all directions to share in the find. Some families in the Fishertown, previously in very humble circumstances, became proud owners of handsome new boats, even houses.

15

ENTERTAINMENT

Every July, a group of pierrots visited the town and set up their stage on the Links. They attempted to charge the audience by setting out chairs within a small fenced area, but of course we always watched from outside, enjoying the amusement free. It always seemed to be wet and cold when the pierrots came. They must have made a very poor living, sending round the collection boxes, and I remember feeling sorry for them because of the small audiences. We became pierrots ourselves. In our big shed we dressed up and performed to each other, singing 'Happy days are here again', the theme song, and dancing and cracking jokes.

The Nairn Games, advertised as 'the only free Games in Scotland', were always held in August and drew in the country people, and also crowds from far and wide who travelled to Nairn by train, bus and car. The arena on the Links, where the Games were held, was open, and it was impossible to charge, except for the small number of people who paid to sit on stands or benches inside a cordoned-off area. Good collections were taken by volunteers in fancy dress, who brought round buckets and a sheet so that people could throw their money from a distance.

A huge fairground was set up to coincide with the Games and was in place a week or two before Games Day. Rents from the owners of the stalls and roundabouts, raised by

Nairn Town Council, were handed over to the Games Committee and this kept them viable. Our great desire was to collect as much money as possible to spend at 'the Showies'. Older relatives were good at handing out sixpences and we were always able to pay for a few visits to our favourite stalls and roundabouts – jungle ride, chairoplanes, bumping cars, helter-skelter, cake-walk, and the ghost train, before prices doubled on Games Day. Most of the Links was covered with stalls and thrilling amusements such as 'Wall of Death', 'Headless Woman' and 'Hall of Mirrors'. I once ventured a visit to the 'Wall of Death', being very curious about it. With some trepidation, I climbed up the outside of the huge wooden cylinder and prepared to watch from the gallery. I was quickly shocked, stunned and frightened by the daring of the motor cyclists speeding round the vertical walls of the cylinder. The noise in the enclosed space was intense because there were at least two motor cycles taking part and they kept revving up to maintain maximum momentum (no doubt too they were adjusted to produce maximum noise for the excitement). Not only did the performers climb the vertical walls on their machines, they actually overtook each other in an incredibly confined space, the only safety provision for the audience being a steel cable about 18 inches below the gallery around which we stood. I couldn't stand it and left almost immediately.

The Globe of Death consisted of a steel lattice sphere. The performance was nearly the same except that, in this case, the motor cyclists actually drove upside down briefly; the audience could certainly feel more secure watching this event because the globe was completely closed.

Near the opening to the fairground, we always stopped to 'Pull the old man's whiskers'. The lady who ran the stall announced appealingly 'All prizes, no blanks!' so we were frequent callers. The lady with crimped white hair stood beside the large face of a man modelled from plaster or papier mâché. Long cords hung down from his mouth, and on payment of threepence (sixpence on Games Day) we pulled one of the cords. Attached, out of sight, to each cord were prizes which we were given. These must have been worthwhile to children as we went back again and again to the stall.

After having our fill of fun before Games Day, we spent what was left of our money on the day itself buying windmills, funny hats, little stuffed birds on sticks, furry monkeys, and Candy Fred's delightful coloured candy. Fred came from Fraserburgh and enjoyed an excellent trade. There were ice-cream vans and food stalls of various kinds to help sustain visitors out for the day. In the wake of the Games and the Fair, we set up our own 'Showies' at the 'Parkie'. Little groups of friends set up 'stalls' and offered 'prizes', sometimes taken from mothers' kitchens, to the winners. Our currency was plain pins and a mark of success was the row of pins, on our jumpers. I expect that we were far less charitable than the real stallholders were.

The Games marked the end of the summer holidays and the season in Nairn. The visitors left, the town suddenly became quiet, and we went back to school for the long term leading up to Christmas. In winter, after homework, our table was set up for playing with Meccano, table tennis or billiards. The boys used towels to make the cushions and pockets for billiards, and many a game was played with the shaft of an old golf club as a cue before they were old enough to go to the Seaman's Hall for a game. We played ludo, tiddlywinks, draughts, snakes and ladders, snap and jigsaws, and the boys inspected their cigarette cards.

Entertainment

Homework took up much of the evenings, so we didn't have much time to fill in. When I was eleven, in 1935, we got our first radio, an Ultra, operated by a big 'dry' and 'wet' battery (an accumulator) which had to be recharged weekly at a local garage for a small fee. We listened avidly to programmes on Children's Hour and this was a welcome addition to our winter entertainment. Before we acquired our own radio we took it in turns with an uncle to use the wireless from the drifter *Glenerne* when it was at home. It was a very primitive instrument with a large separate loudspeaker set on the top of it.

The beach, of course, was a wonderful summer playground. From an early age we were aware of the dangers of the sea and were merely told to watch ourselves when we set off daily with little brothers and sisters to wade, dig and build sandcastles. This long therapy of water and sand meant we were usually more than ready to start school again at the end of the holidays.

16

FOOD

Several times a week my mother made girdle scones and pancakes. We arrived home from school just in time to get one or two warm, before going out to play. Dinner in the middle of the day and tea at five were the main meals. For dinner there was usually hot soup, summer and winter, followed by meat from the stock, and then dumplings, which were steamed suet puddings containing dried fruit, apples or golden syrup, with custard. We were well fortified for afternoon school and had a good tea in the evening. There would be bacon and eggs, fish, potted meat or meat roll with fried potatoes, scones and pancakes with jam. The meat roll was made with mince, bacon, breadcrumbs and eggs and steamed in a stone jar in a pan of boiling water. The fried potatoes were left-over mashed ones, cut into slices and fried. We called them cheevies. A slice of bread was also called a cheeve, mainly by older people. In season we ate a lot of cod and haddock roe. The soups my mother made were Scotch broth, lentil, potato, mince and fish, and, on very rare occasions, chicken.

Mince soup with potatoes in it was a complete meal. I used to love 'overdays' soup as it was thicker and tastier than on the first day. The frying pan was used a lot at teatime. My mother sometimes made pancakes the size of the frying pan which she then tossed and caught in the pan for our amusement. I never saw one fall on the floor.

Food

Stock for broth was made with brisket, flank of beef or mutton which gave it a wonderful flavour, and was very tasty eaten cold with potatoes, turnip and carrots, served in our empty soup plates. I remember helping to dice the carrots and turnips on Saturday night for Sunday's broth. Meat and root vegetables featured prominently in our diet, as of course did fish of many kinds and best quality. We ate a lot of fish and shellfish such as prawns, crabs and cockles. A few people gathered and sold cockles at the doors. They could be gathered from the East Beach, well away from the risk of pollution. Before World War II, my father and his brother, fishing out of Nairn in a motor boat aptly named *The Brothers*, used to catch large and beautiful prawns which were not in great demand at the time, so we ate many of them. Left-over white fish were flaked and made into fish cakes with mashed potato and egg. On Saturdays we ate salt herrings with potatoes. The fishermen brought home small barrels of herring to help feed their families in winter, and our custom was to have them on Saturdays when we were home from school. We ate several at dinnertime without, in my case, really enjoying them. As they were just boiled and served with potatoes, it was an easy meal, giving mothers a break. Sometimes on Saturdays, if there was money available, we had meat pies, small individual ones, still favoured by Scots people and baked locally. On Sundays, in wintertime, when the fire was at a suitable stage, we cooked dried fish on a brander over the red ashes. The skin blistered and burnt a little, but this made them tastier and we ate them with our fingers, pulling the flesh away from the bone. We dried flat fish mainly, plaice or greybacks (dabs or grey sole). They were skewered on a spit and dried outside for a few days or sometimes hung up in the range away from the smoke of the coal.

Stewed meat, mince, rabbit, and sometimes steak pies, were the mainstays of our diet, along with baked or steamed rice and other milk puddings and baked apples or rhubarb sponge, and stewed apples with rich tea biscuits crumbled over them. My mother hated the smell of rabbit cooking, but we loved the taste of its rich gravy. Rabbits were used a lot because they were cheap. We bought them from the butcher, ready skinned. Sometimes we bought them at the door from a man who trapped them.

Fish soup (we called it fish sauce) was made from whole haddocks. Fish on the bone gave the best flavour and the other ingredients were chopped potatoes, onions, milk, butter and seasoning. When mince was being stewed, dough boys (flour and suet balls) were added twenty minutes before the end of cooking time and this made the meal quite substantial.

Salads were alien to us until much later, probably because the vegetables needed were either unavailable or too expensive, and no use for keeping out the cold. My mother encouraged us to eat porridge for breakfast, but we weren't keen and it used to bring Johnny out in heat spots. He ate brose made with pease meal mixed with boiling water and seasoned. Corn flakes, Grape Nuts and other cereals were then freely available and we ate these for breakfast, along with butter rolls and white morning rolls from Dodo's, a baker just up the road. Dodo's butter rolls were as delicate and tasty as the finest French croissants. He also made lovely teabread which we had as a treat from time to time.

During the summer, when eggs were plentiful and cheap, we preserved a good few dozen in waterglass for using in the winter. The preservative was mixed with water and kept in zinc pails. The eggs were gently laid in this in layers and kept in a dark cupboard. When needed, we were sent to get them and hated having to put our hands into the ice-cold water-glass. The eggs preserved very well and were useful for baking in the winter months. We never ate them scrambled, boiled or fried, although they were probably quite safe.

Jam and marmalade were always made in their seasons and saw us through the whole year. Marmalade was made, as it is today, from Seville oranges. Mr Davidson, who had a fruit-erer's shop at the corner of High Street and Gordon Street, lent out to his customers a machine for slicing the fruit. We were given the job of operating the machine and we also helped to cover and label the jars. In the summertime we gathered rasps from open scrub areas and quarries around the town, and in autumn, brambles. Both jam and jelly were made from brambles. The berries were first cooked until soft, then strained through muslin; the liquid was boiled up with sugar to setting point. Blackcurrants, strawberries, rhubarb, gooseberries and apples were also used to make jams and jellies, but I think we stuck to those that were free, either as gifts from friends' gardens or those we picked ourselves. We ate a lot of bread and jam when we came in hungry from play. Chocolate biscuits were unknown to us.

In the absence of any form of refrigeration, fresh food was bought daily. Our shed faced north and was used quite a lot for keeping things cool. Milk bottles were stood in bowls of water and butter was also kept in water in the summer time. Our butcher sent the meat by message boy and our milk was delivered in the early hours of the day by a cart and horse driven by Allan Leith.

APPENDIX I

NAIRN IN THE MORAY FIRTH (SOURCE UNKNOWN)

Fishing villages were established all along the Moray Firth coast. Burghead about 16 miles to the east of Nairn was the principal Norse settlement where a Fire Festival called the 'Burning of the Clavie' continues to the present day.

The principal name was 'Main' which is regarded as being the anglicised version of the Norse 'Magnus'. So much confusion was caused among strangers by the great number of Mains that to differentiate between them they adopted tee-names or family names such as Ellen, Bochel, Duggie, Sailor, Mein, Ian and Bunker. A number of them adopted the clan name Mackintosh because of the proximity of the original clan land to Nairn. The Parish of Petty extended east to the old village of the Fishertown of Delnies, where the Mackintosh's and Wallaces came from, to Nairn after the building of the harbour in 1820.

Fishing continued and prospered in the Moray Firth using small fishing boats of a size that could be pulled up above high water mark on the beaches by two or three men. It was not until the harbour was constructed that larger boats were built which enabled the fishermen to extend their fishing grounds beyond the inner firth to the Orkney and Shetland Isles and through the Pentland Firth to the west coast.

Food

The most prosperous time in the fishing industry came with the advent of the steam drifter about the turn of the century (the first came to Nairn in 1902) and continued until World War I when the German and Russian markets for salt herring were lost. The industry flourished for a short time after the war and then slowly declined until the present day, when no fishing at all takes place from the harbour.

Appendix 2

Culbin Sands

The stretch of shoreline in the Moray Firth from Whiteness Head in the west to Findhorn in the east had been for centuries unstable due to strong winds blowing from the west and nor'east.

Bent grass from the shore was used extensively by the fisherfolk for thatching houses and for lining fish creels and scoos [shallow baskets for holding fishing lines], and it took a large amount of Bent grass to thatch one house. This continual plundering of the shore made the sand unstable and the resultant sandstorms caused the river estuary to silt up and change course. The westerly winds pushed the river east to below Kingsteps and an easterly sandstorm pushed the river west to enter the sea at Nairn.

In October 1694, a strong wind from the west continued to blow sand eastwards for a period of three days' blocking the estuary of the river Findhorn and changing its course to the east, away from the Old Bar. The estate of Culbin which was regarded as the granary of Moray and comprised 16 farms, several smallholdings, a mansion house with a large orchard, a church and farm cottages for estate workers and their families, was engulfed and completely destroyed.

The storm came suddenly and without warning, taking the entire population of Culbin and the fishing village of Maviston by surprise. A ploughman had to unshackle his horse

and leave his plough in the field, and reapers in a field of late
barley had to run for their lives. In a few hours, the plough
and the barley were buried beneath the sand and when the
storm blew itself out 3,600 acres of land were lost forever.

Parts of buildings were uncovered for short periods over
the years until the Forestry Commission bought the land in
1946 and stabilised the sand by planting trees.[1]

1 For a recent study, see Sinclair Ross, *The Culbin Sands – Fact and
Fiction*, Centre for Scottish Studies, Aberdeen, 1992.

GLOSSARY

Alease	Variant of the children's game hide and seek.
Bannies	Sticklebacks.
Beddies	A local childhood nickname for hopscotch (see also Skeetchies).
Bidding	Wedding invitation.
Boolies	Children's marbles.
Bow	Buoy used by fishermen.
Box-bed	Bed enclosed in a wooden panelling, the front having either sliding panels, hinged doors, or curtains.
Brander	Gridiron used over a fire for cooking.
Bullace, bullax	Hand-held axe or hatchet.
Bush, bush rope	Rope to which drift nets are attached.
Caff, cauff, calf	Chaff, often used for stuffing mattresses etc.
Caff-seck	A sack full of chaff.
Cawed	Driven.
Cheevies	Potatoes boiled, sliced, then fried.
Churls, chirls	Kindling wood.
Contrack	Ceremony whereby families witnessed and approved marriage betrothal. A precursor of formalised wedding arrangements.

Glossary

Crowdie	Soft cheese.
Cruisie, croozie	Open, boat-shaped lamp with rush wick.
Dolly	Wooden shaft attached to a disk with projecting arms, used for beating and stirring clothes in a washing-tub.
Donal	A larger version of a marble (see Peachie).
Durkin	A fir cone.
Eye-aac	A local childhood nickname for a version of marbles.
Farlin, farlan	Long box or trough in fish-curing yards into which herrings were emptied for gutting.
Freestone, freestane	Block of easily worked sandstone.
Fulls	Herring full of milt or roe.
Gird	Metal hoop bowled along with a stick.
Girdle, gridle, griddle	A flat, iron plate, with a low handle, for baking oatcakes, scones and pancakes above the fire.
Glassie	A glass marble.
Groyne	Breakwater, of wood or other material, to check erosion and sand-drifting.
Guernsey	Close-fitting upper garment, a jersey.
Kick the tinnie	Variant of the children's game, hide and seek.
Kirkin', kirking	First church attendance by a newly married couple.
Kist	Chest, large box.
Knock, nock	Clock.

Salt Herring on Saturday

Lug worm, lugworm
                    Worm found in the sand at the
                    seashore, used for bait.
Maister   Only or principal teacher in a small
                    rural community.
Marram grass   Seaside grass, binder of sand dunes.
Matt full   Sexually mature herring.
Matties   Young female herring with the roe not
                    fully developed.
Melg   Soft roe of male fish.
Mizzen sail   The hindmost of the fore and aft sails.
Mortcloth   Pall, cloth covering a coffin.
Mouly heels   Chilblains, especially broken ones on the
                    heel.
Peachie   A smaller version of the childhood toy,
                    a marble (see Donal).
Piles   Wooden groynes used to restrict
                    erosion.
Ran, rawn   Fish roe.
Range   Enclosed, metal kitchen fireplace.
Redd   Disentangle, e.g. fishing lines.
Scaup   Thin shallow soil, an infertile piece of
                    stony ground, filled with water.
Scoo/sku   A scull, a shallow, scoop-shaped oval
                    basket for fishing lines etc.
Scrougened,   Partly dried.
   scroonged*
Shottle   Small compartment inside a trunk, chest
                    etc.
Skaillet, skillet Small, long-handled pan.
Skeetchies   A local childhood nickname for
                    hopscotch (see also Beddies).

Glossary

Skuttlies	Childhood term for narrow alleys.
Smas	Immature fish.
Stonies	Marbles made of ceramic rather than glass.
Strannies	Drainage gullies at the roadside.
Thrum rug	Small carpet made from the ends of warp-thread.
Tinter, tenter	Bar of wood fitted with hooks on which fish are hung to dry.
Tippen	The horsehair cords for the hooks of a fishing-line.
Weers, wires	Steel knitting needles.
Worsit, wirsit	Worsted.
Zulu	Type of fishing boat common especially in the Moray Firth and the Clyde from *c*.1880-1905, with fore- and mizzen masts, straight prow and markedly raked stern.